Functional Biology of Plants

Functional Biology of Plants

Editor

Raj Hulasare

scitus
academics

Functional Biology of Plants

Edited by **Raj Hulasare**

ISBN: 978-1-68117-065-7

Library of Congress Control Number: 2015935463

SCITUS Academics LLC,
www.scitusacademics.com
Box No. 4766, 616 Corporate Way,
Suite 2, Valley Cottage,
NY 10989

Printed in United States of America on Acid Free Paper ∞

Contents

Preface

The department of plant science sees how plants live and function. The ultimate goal is usually to explain all life processes of plant life through a small number of comprehensive chemistry, mathematics and physics principles. Plant physiology aims at understanding all aspects of plant life. Fundamental processes such as photosynthesis, respiration, plant nutrition, plant hormone functions, tropisms, nasal movements, photomorphogenesis, circadian rhythms, environmental stress physiology, seed germination, dormancy and stomata function and transpiration are studied by plant physiologists in both parts of the plant water relationship. All living things, like plants, need constant energy to maintain and repair their highly organized structures and to grow and reproduce. Chemical potential is a quantitative expression of a substance's free energy. Water-related free energy is represented by the chemical potential of the water. Drinking water flows from areas with a higher chemical potential to areas with a lower chemical potential without energy input.

This book provides a contemporary, applied integration of molecular and plant-wide approaches for plant study. The book also addresses the release, transformation and absorption of materials and their movement within and between the plant's cells and organs; the physiology of growth, development and reproduction, which deals with these elements of plant function.

Chapter 1

Expression-Based Network Biology Identifies Immune-Related Functional Modules Involved in Plant Defense

Joel P Tully[1], Aubrey E Hill[1], Hadia MR Ahmed[1], Ryan Whitley[1], Anthony Skjellum[1, 2], and M Shahid Mukhtar[3, 4]

[1]Department of Computer & Information Sciences, University of Alabama at Birmingham, Birmingham, AL, 35294-1170, USA

[2]Current Address: Samuel Ginn College of Engineering, 3101 Shelby Center, Auburn University, Auburn, AL, 36849-5347, USA

[3]Department of Biology, University of Alabama at Birmingham, Birmingham, AL, 35294-1170, USA

[4]Nutrition Obesity Research Center, University of Alabama at Birmingham, Birmingham, AL, 35294, USA

ABSTRACT

Background

Plants respond to diverse environmental cues including microbial perturbations by coordinated regulation of thousands of genes. These intricate transcriptional regulatory interactions depend on the recognition of specific promoter sequences by regulatory transcription factors. The combinatorial and cooperative action of multiple transcription factors defines a regulatory network that enables plant cells to respond to distinct biological signals. The identification of immune-related modules in large-scale transcriptional regulatory networks can reveal the mechanisms by which exposure to a pathogen elicits a precise phenotypic immune response.

Results

We have generated a large-scale immune co-expression network using a comprehensive set of*Arabidopsis thaliana* (hereafter Arabidopsis) transcriptomic data, which consists of a wide spectrum of immune responses to pathogens or pathogen-mimicking stimuli treatments. We employed both linear and non-linear models to generate Arabidopsis immune co-expression regulatory (AICR) network. We computed network topological properties and ascertained that this newly constructed immune network is densely connected, possesses hubs, exhibits high modularity, and displays hallmarks of a "real" biological network. We partitioned the network and identified 156 novel modules related to immune functions. Gene Ontology (GO) enrichment analyses provided insight into the key biological processes involved in determining finely tuned immune responses. We also developed novel software called OCCEAN (*O*ne *C*lick *Cis*-regulatory *E*lements *AN*alysis) to discover statistically enriched promoter elements in the upstream regulatory regions of Arabidopsis at a whole genome level. We demonstrated that OCCEAN exhibits higher precision than the existing promoter element discovery tools. In light of known and newly discovered *cis*-regulatory elements, we evaluated biological significance of two key immune-related functional modules and proposed mechanism(s) to explain how large sets of diverse GO genes coherently function to mount effective immune responses.

Conclusions

We used a network-based, top-down approach to discover immune-related modules from transcriptomic data in Arabidopsis. Detailed analyses of these functional modules reveal new insight into the topological properties of immune co-expression networks and a comprehensive understanding of multifaceted plant defense responses. We present evidence that our newly developed software, OCCEAN, could become a popular tool for the Arabidopsis research community as well as potentially expand to analyze other eukaryotic genomes

BACKGROUND

Agriculture, in struggling to meet global food demands of a rapidly growing population, has been plagued with plant diseases that thwart crop production and account for a global annual average yield loss of 16 percent [1-3]. In light of the importance of agriculture to humans, it is essential to elucidate and understand the mechanisms of plant immunity at the molecular level. As with any host-pathogen conflict, plants and their disease agents are in an evolutionary arms race. When the host mounts a defense reaction, the pathogen develops new strategies to evade the defensive mechanisms, which causes the continuation of this cycle, *ad infinitum*[4,5]. Our present understanding of plant immune systems reveals two primary means by which plants recognize their invaders. The recognition of Microbial-Associated Molecular Patterns (MAMPs) is the first line of defense and its associated MAMPs-Triggered Immunity (MTI) is highly efficient at repelling most pathogens. If pathogens are able to breach this first line of defense through the production of effector proteins, plants deploy Effector-Triggered Immunity (ETI) [4,6]. Both modes of defense cause massive transcriptional reprogramming which involves complex signal transduction networks and cross-talk that is mediated by plant hormones, including: Salicylic Acid (SA), Jasmonic Acid (JA) and Ethylene (ET) [7,8]. The Arabidopsis genome encodes approximately 1,922 transcription factors (TFs) that are implicated in diverse biological processes including the regulation of immune signaling pathways [9-11]. However, the global organization of TF-DNA interactions remains elusive. Moreover, what remains mostly unknown are the precise mechanisms by which the plant cell integrates innumerable synergistic and antagonistic immune transcriptomic signals to orchestrate fine-tuned and pathogen-specific defense responses.

In recent years, systems biology approaches, specifically network analyses that integrate experimentally derived information with computational modeling, have emerged as powerful tools for studying complex traits in diverse species [5,12-15]. Network biology provides comprehensive analyses of the system's components (nodes) and the relationships among them (edges). In addition, the analyses of the topological properties of the network can provide further understanding of the hierarchical organization of a complex biological system and contribute to the overall interpretation of biological complexity. In any eukaryotic cell, thousands of genes and their products orchestrate their transcriptional and translational activities to ensure the proper execution of cellular functions[13,15,16]. It has become evident that biological functions can be accomplished by functional modules that are embedded within the interaction networks including transcriptional gene regulatory networks [17,18]. Comprehensive high-throughput analyses of gene expression can allow for the identification of gene clusters that are highly correlated in expression levels across multiple samples in any given cellular state [13,16]. Generally, it is thought that genes in the same co-expression sub-network are often enriched with similar functional annotations. Additionally, the metric to measure the co-expression falls into one of two major categories: correlation coefficients or mutual information measures [19]. Finally, finding common *cis*-regulatory elements (transcription factor binding sites) can aid in the identification of co-regulated gene clusters and characterization of transcriptional regulatory networks [13,14,17].

In the current study we employed a systems biology approach and report the construction of a large-scale Arabidopsis immune co-expression regulatory (AICR) network. We tested five diverse algorithms, *i.e.* PCC (Pearson Correlation Coefficient), ARACNE multiplicative (Algorithm for the Reconstruction of Accurate Cellular Networks), ARACNE additive, CLR (Context Likelihood of Relatedness), and MRNET (Minimum Redundancy NETwork) with different thresholds, which yielded 15 pairs of experimental networks along with their respective random networks [19-25]. We employed network biology analyses and determined that ARACNE multiplicative network (5,147 nodes and 38,610 edges), with threshold of 0.8 exhibits properties of a "true" network as it possesses the scale-freeness attribute (degree distribution follows a power law). Next, we partitioned the AICR network and predicted 156 functional modules containing at least six members with the largest module encompassing 178 nodes. Subsequently, we analyzed functional annotations of genes within each module and calculated enrichment of specific Gene Ontology terms to evaluate the biological significance of functional modules. To establish a causal relationship between co-expression and co-regulation, we sought to identify common *cis*-regulatory elements. First we developed a new, comprehensive software interface for *cis*-regulatory elements discovery in Arabidopsis (named OCCEAN - One Click *Cis*-regulatory Elements ANalysis). We demonstrated that OCCEAN boasts higher capacity (can process the entire genome scale; over 30 million characters in a single run) and features higher precision than MEME (Multiple EM for Motif Elicitation). We identified several statistically enriched novel *cis*-regulatory elements in our dataset. Finally, we evaluated and discussed key immune-related functional modules.

RESULTS AND DISCUSSION

Construction of the Arabidopsis Immune Co-Expression Regulatory (AICR) Network

To generate a large-scale immune co-expression network, we selected Arabidopsis transcriptomic data that encompassed the broadest possible spectrum of immune responses to pathogens or pathogen-mimicking stimuli treatments [26-34]. Previously, we have employed the same set of experiments to define transcriptional responses of genes that encode the proteins identified in plant-pathogen immune network, version 1' (PPIN-1) [5] We compiled the lists of probes showing significant up- or down-regulation and discovered 8,377 differentially expressed genes, which subsequently were used to build a comprehensive immune co-expression network In a co-expression network, the nodes represent the genes and the edges (lines connecting nodes) represent the similarity/relatedness between the genes, which is generally measured by the Pearson correlation coefficient (PCC) [18]. PCC and other measures of association determine the degree of linear dependencies between the variables [19]. While these association models have obvious statistical advantages, they lack the ability to capture non-linear relationships [35]. In contrast, mutual information (MI) is a non-linear statistic that provides an attractive alternative to measure biologically significant non-linear relationships [36]. We used both linear and non-linear models to generate Arabidopsis immune co-expression regulatory (AICR) network. To measure linear associations between two genes, we calculated the PCC values for all of the genes in a matrix of $8{,}377 \times 8{,}377$ (over 70 million combinations). Then, we employed both a similarity threshold algorithm (a predefined single cut-off value) as well as a newly developed and parameter-free algorithm termed mutual *k*-nearest neighbor (mKNN) to process the calculated PCC values [18]. mKNN was recently shown to possess several advantages over the commonly used similarity threshold algorithm [18]. Using the linear model, we generated seven pairs of experimental networks along with their respective random networks, *i.e.*, PCC (0.9), K3, K10, K20, K50, K100 and K250 (Table 1). To infer non-linear association between immune-related genes, we employed four recently used MI methods: ARACNE multiplicative (Algorithm for the Reconstruction of Accurate Cellular Networks), ARACNE additive, CLR (Context Likelihood of Relatedness), and MRNET (Minimum Redundancy NETwork) with 0.8 and 0.9 thresholds. Specifically, we employed the 'parmigene' package (PARallel Mutual Information estimation for GEne NEtwork reconstruction, an R package) to construct eight Additional experimental networks [21-25]. It has been recently shown that the MI estimator implemented in parmigene provides more precision and unbiased results compared with previous MI estimators (Figure 1). Our approach to initially build multiple experimental and random networks using both linear and non-linear models is aimed to determine an optimal experimental network that displays topological properties of a "real" biological network [15]

Table 1: Comparison of multiple algorithms used to generate Arabidopsis Immune Co-expression (AICR) Network

Algorithm	Threshold	Nodes	Edges
PCC K=3	Top 3	8250	1371822
PCC K=10	Top 10	8336	1374677
PCC K=20	Top 20	8358	1379968
PCC K=50	Top 50	8369	1396874
PCC K=100	Top 100	8373	1453118
PCC K=250	Top 250	8377	1758775
PCC	90.00%	8377	2755191
ARACNE additive	90.00%	1999	10253
ARACNE multiplicative	90.00%	1999	10253
MRNET	90.00%	1999	3493
CLR	90.00%	12	11
ARACNE additive	80.00%	5147	32972
ARACNE multiplicative	80.00%	5147	38610
MRNET	80.00%	5058	6049
CLR	80.00%	57	60

PCC, pearson correlation coefficients; *ARACNE*, algorithm for the reconstruction of accurate cellular networks; *CLR*, context likelihood of relatedness; *MRNET*, minimum redundancy NETwork.

Tullyet *al.*

Tullyet *al. BMC Genomics* 2014 **15**:421 doi:10.1186/1471-2164-15-421

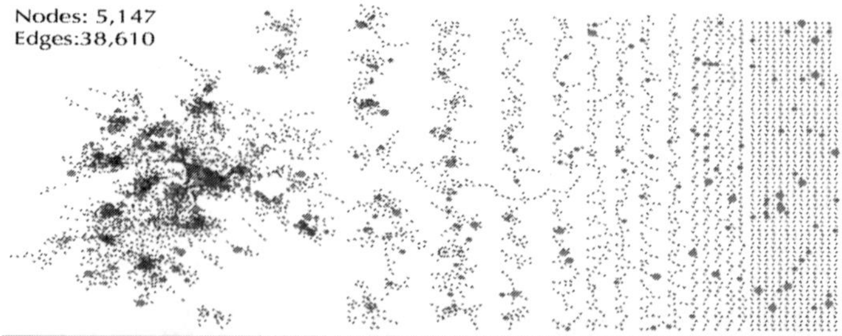

Network	Clustering coefficient	Connected components	Network diameter	Network density	Network centralization	A verage number of neighbors
ARA CNE multiplica tive (0.8 thre shold)	0.306	759	34	0.001	0.022	7.501

Node — Edge PPIN-1 node PPIN-1 node and effector target

Figure 1: Construction of Arabidopsis immune co-expression regulatory (AICR) network. Network is displayed in Prefuse Force Directed Layout algorithm in Cytoscape. A node (blue circle) represents genes and a grey edge linking two nodes indicates the co-expression relationship between these two nodes based on Pearson Correlation Coefficient (PCC). Red nodes represent the proteins present in Plant-pathogen Interaction network (PPIN-1) [5]. The common feature of experimental the AICR network is listed on the bottom.

Global Topological Properties of the AICR Network

Network topology refers to the arrangement or pattern of interactions within a network. The majority of naturally occurring networks, including biological networks, maintain certain topological characteristics in terms of their structure and organization that are significantly different from random networks [15,37,38]. Therefore, in an initial experiment, we computed degree of distribution for all 15 pairs of experimental networks along with their respective random networks. We found that two networks, ARACNE multiplicative (5,147 nodes and 38,610 edges) and ARACNE additive (5,147 nodes and 32,972 edges), with threshold of 0.8 exhibit the scale-freeness attribute (degree distribution follows a power law), when compared with all random networks and other experimental networks (Figure 2). Since the precision of parmigene-based ARACNE multiplicative is better than ARACNE additive, we selected ARACNE multiplicative with threshold 0.8 as the "true" experimental AICR network [25].

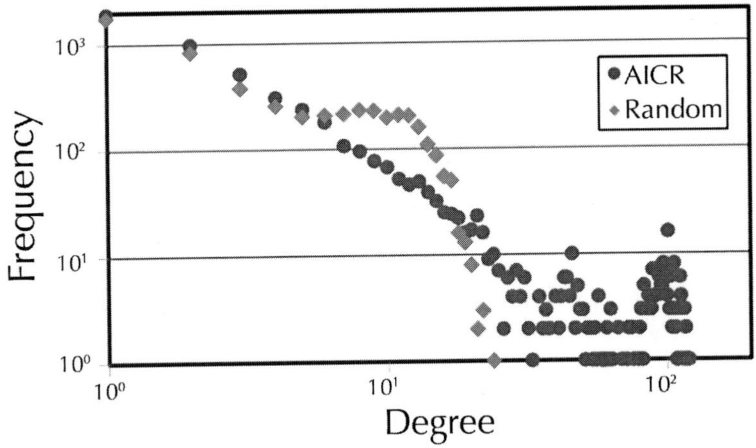

Figure 2: Degree distribution of the AICR. Frequency of the degree of the AICR (blue circles) and random (red diamonds) networks are indicated in log scale. Presence of highly connected nodes (hubs) can be observed in the AICR network.

"Real-world networks" tend to form high-density clusters and exhibit a clustering coefficient that is significantly higher than expected by random chance [14,18]. Therefore, we computed clustering coefficients of the AICR network and its corresponding random network (Figure 3). A clustering coefficient describes the degree of congregation among the nodes of a graph. The distribution of the average clustering coefficient in the AICR ranges between 0.4-0.8 and the frequency of the number of neighbors with higher clustering coefficient in the AICR is significantly greater than in a random network (Figure 3). To understand how the differences in the number of connections impact the topology of our co-expression network, we measured the shortest path (shortest distance between all pairs of nodes) and the closeness centrality (the inverse sum of shortest distances to all other nodes from a focal node) [39]. The path lengths for the majority of the nodes in the AICR are significantly shorter than those of a random network. In addition, the number of neighbors with significantly higher closeness centrality (threshold 0.2) is higher in the AICR network compared with random network. In biological networks, highly connected nodes (hubs) and nodes central to the network (betweenness) are thought to play significant regulatory roles on their adjacent nodes. Thus, we computed several topological properties including shared neighbor distribution (number of interaction partners shared between two nodes), neighborhood connectivity distribution (average connectivity of all neighbors), topological coefficients (the tendency of the nodes in the network to have shared neighbors) and betweenness centrality (nodes' centrality in a network) (Figure 4)[37,38]. These data suggest that the AICR network is not randomly organized and shares properties of several previously described biological networks [5,12-15]. Another important characteristic of scale-free networks is the presence of a main component. Connected component analyses discovered that the AICR network comprises 63 disconnected components, each containing at least six members. The main component in the AICR network has 2379 nodes and 33214 (86%) edges. The second largest component contains only 79 nodes and 310 edges. This qualitative global topology of the AICR network is similar to previously known biological networks in Arabidopsis and other eukaryotes[14,40]. In addition, network biology analyses revealed that degree distribution of main component of the AICR follows a power law compared to the main component of a random network with similar size

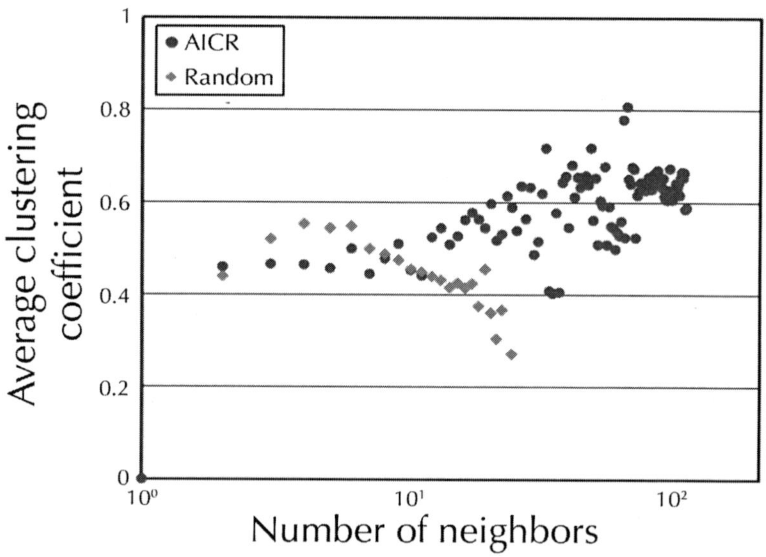

Figure 3: Average clustering coefficient of the AICR. A plot between number of neighbors (X-axis) and average clustering coefficient (Y-axis) is drawn representing the AICR (blue circles) and random (red diamonds) networks.

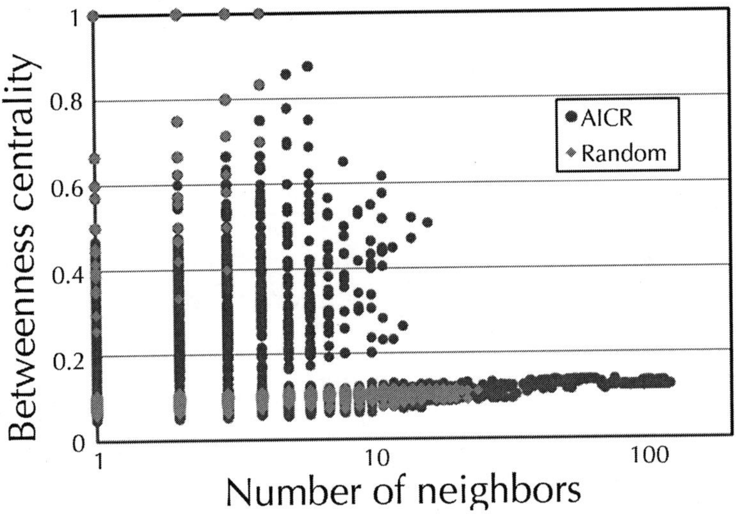

Figure 4: Betweenness centrality of the AICR. Frequency of Betweenness centrality of the AICR (blue circles) and random (red diamonds) networks are indicated.

Biological Implications of Topological Properties of the AICR Network

Topological properties of the network can also be utilized to prioritize genes for further functional analysis. Highly connected and highly centered nodes are likely to play crucial roles in maintaining network integrity and controlling information flow throughout the plant immune system. Specifically, we identified 190 highly connected nodes (Hub^{50} encompassing at least 50 connections in the AICR). To characterize the significance of these highly connected proteins, we combined Hub^{50} of the AICR with biophysical interactions from plant-pathogen immune network, version 1′ (PPIN-1; Figure 1) [5]. The PPIN-1 was constructed using effectors (virulence factors) from two pathogens, Arabidopsis immune proteins, and ~8,000 other Arabidopsis proteins covering almost one-third of the genome[5]. Pathogens utilize effector molecules to rewire the host network in a manner conducive to pathogen proliferation and dispersal [41]. PPIN-1 discovered 165 effector interacting proteins (effector targets) and it was also demonstrated that these effector targets are highly interconnected host proteins [5]. Despite the unavailability of effector targets from the entire Arabidopsis genome in PPIN-1, we found that the AICR network contains 51 effector targets and three proteins (At3g63210, At4g11890 and At3g07780) are Hub[50] in the AICR network (Figure 1). Given that transcriptional regulation does not necessarily coincide with biophysical interactions, this overlap is potentially significant and biologically meaningful. This also suggests that pathogen effectors target key regulatory proteins in both transcriptional as well as protein-protein interactions networks.

In addition, we selected ninety nodes ("top 90") with the highest betweeness centrality (Figure 4). Among them, we found a number of previously characterized immune regulators or defense-related proteins, encompassing various levels of signal transduction flow: resistance proteins (RPP4, uncharacterized TIR-NBS-LRR) [42,43], kinases mediating early signaling events (MAPKKK13, WAK) [44], transcriptional executors (ANAC072, WRKY54, MYB66, WER1) [45-47], secretory proteins assisting with folding and modification of newly synthesized proteins (TRAP, CNX3, Sec14p, cyclophilin, UDP-Glycosyltransferase, DnaJ family, KMS1, GlcNAc1pUT2) [48,49], enzymes involved in production of phytohormones and antimicrobial compounds (AOC2, ILL3, PAL2) and pathogenesis-related proteins (PR1) [50,51], hormonal response modulators (PYL6, IAA28) [52,53] and finally, components of ubiquitin-mediated protein degradation (UBQ4, an F-box protein that interacts with SKP1) [54,55]. Identification of these key immune regulators serves as a proof of concept for our analytical approach and justifies further analyses on suite of "top 90" proteins selected among thousands of other proteins in the AICR network.

Network Clustering and Module Annotations

Another essential feature of a majority of biological networks is their ability to naturally organize into modules [14,17,18]. Discovery of such functional modules within biological networks is imperative for understanding principles of cellular organization and functions. To identify functional modules (subnetworks) that are comprised of highly interconnected sets of genes within the AICR network, we employed a fast agglomerative algorithm (FAG-EC) [56]. This new, low time complexity algorithm operates based on a local variable, edge clustering coefficients, making it ideal to partition a relatively large network [37,56]. Overall, we identified 156 immune-related functional modules containing at least six members with the largest module encompassing 178 nodes Furthermore, we subjected the AICR network to a module size distribution analysis. We computed the frequency and module size and demonstrated that the distribution of the module size follows a power law distribution with an exponential truncation, which is common for several previously described "real-world" networks [14,18]. The module size distribution property further indicates that the AICR network possesses a modular structure.

We also investigated the enrichment of Gene Ontology (GO) terms [57,58] in the ten largest immune-related modules ("top 10") with sizes ranging from 38 to 178 distinct nodes using the Database for Annotation, Visualization and Integrated Discovery (DAVID) v6.7 [59] The GO enrichment terms are remarkably informative in examining significant biological functions among co-expressed genes in large-scale networks [58,59]. Several gene ontology (GO) categories were enriched (p-values >0.01) among the genes in the AICR network (Table 2). These major GO categories include: immune-related, plastid, reproductive processes, nucleotide binding/kinases, intrinsic to membrane, metal-binding, mitochondrion and non-membrane-bounded organelle. We utilized these GO enrichments in combination with significantly enriched *cis*-regulatory elements to define the potential functions of the immune-related modules (described below).

Table 2: Significantly enriched GO terms in the ten largest immune-related modules with sizes ranging from 38 to 178 genes

Module	Go term	Count	%	P value	**Fold enrichment**
1	GO:0000166 ~ nucleotide binding	46	25.84	6.946003862092387E-5	1.732
1	GO:0006796 ~ phosphate metabolic process	31	17.42	5.723269851412882E-7	2.705
1	GO:0004672 ~ protein kinase activity	29	16.29	2.6741028021510666E-8	3.278
1	GO:0008219 ~ cell death	14	7.87	4.28780370877740953E-7	6.123
2	GO:0009536 ~ plastid	31	30.69	9.044388056227762E-4	1.748
2	GO:0004672 ~ protein kinase activity	12	11.88	0.007513	2.463
2	GO:0019748 ~ secondary metabolic process	8	7.92	0.003083	4.073
3	GO:0009536 ~ plastid	36	49.32	3.047542263866877E-10	2.718
3	GO:0055114 ~ oxidation reduction	11	15.07	0.002191	3.019
4	GO:0005886 ~ plasma membrane	16	22.22	0.001229	2.356
4	GO:0006793 ~ phosphorus metabolic process	14	19.44	9.785189944953485E-5	3.418
5	GO:0009628 ~ response to abiotic stimulus	9	15.52	0.009708	2.845

5	GO:0005618 ~ cell wall	5	8.62	0.056662	3.316
6	GO:0009507 ~ chloroplast	22	44.00	1.0752825917609916E-5	2.499
6	GO:0051186 ~ cofactor metabolic process	7	14.00	2.8760079199003342E-5	10.975
7	GO:0000166 ~ nucleotide binding	12	25.53	0.044134	1.777
8	GO:0009725 ~ response to hormone stimulus	8	17.39	0.001339	4.400
8	GO:0006793 ~ phosphorus metabolic process	6	13.04	0.093212	2.354
9	GO:0031224 ~ intrinsic to membrane	13	33.33	0.002493	2.407
9	GO:0006793 ~ phosphorus metabolic process	8	20.51	0.008887	3.139
10	GO:0032555 ~ purine ribonucleotide binding	10	26.32	0.083444	1.765
10	GO:0005886 ~ plasma membrane	8	21.05	0.052468	2.151

Tullyet *al.*

Tully *et al. BMC Genomics*2014**15**:421 doi:10.1186/1471-2164-15-421

Development of OCCEAN Software and Identification of *Cis*-Regulatory Elements

We sought to investigate whether all or a subset of the co-expressed genes within a given immune-related module are also co-regulated (*i.e.,* are direct targets of a common transcription factor). Given that transcriptional regulatory networks are highly complex and that functional modules may display crosstalk among themselves, we can also expect that the same transcription factor can regulate co-expressed genes in multiple immune-related modules. Whereas experimentally verified plant *cis*-regulatory elements can be retrieved from PLACE (A Database of Plant *Cis*-acting Regulatory DNA Elements) [60], Athena (Arabidopsis thaliana expression network analysis) [61] and AGRIS (The Arabidopsis Gene Regulatory Information Server) [11,62], only ~120 *cis*-regulatory elements are currently known in Arabidopsis, and a limited set of ~30 immune-related *cis*-elements have been described. In addition, the prediction of *cis*-elements from classical approaches is typically driven by a single experiment or dataset. Moreover, the currently available online motif discovery software has limited capacity to process a small number of gene entries. For example, frequently used software MEME (Multiple EM for Motif Elicitation) can only process up to 60,000 characters in a single run (*e.g.,*60 promoters, each of 1,000 bp in length) [63,64]. Another common software used to predict *cis*-regulatory elements, AthaMap (Arabidopsis thaliana Map) can only manage up to 200 gene entries[65]. These limitations can substantially decrease the chance of identifying novel *cis*-regulatory elements. Thus, we first aimed to develop OCCEAN (*O*ne-*C*lick *C*is-regulatory *E*lement *AN*alysis) software that can process the promoter gene sequences at the entire genome scale (over 30 million characters) in a single run. OCCEAN identifies the statistically enriched/depleted *cis*-regulatory elements (Figure 5) and integrates data from (*i*) our newly developed BLASTN-based program to identify short sequences and (*ii*) an improved version of the bootstrapping tool POBO (a promoter bootstrapping program) [64]. This user-friendly software requires the sequences in FASTA format as input, processes genome-scale level sequences, identifies common sequences in a given set of promoters, performs bootstrapping, and provides statistically enriched *cis*-regulatory elements in the gene's promoter as the output. OCCEAN is freely available online athttp://occean.cis.uab.edu:8080/occean. We employed OCCEAN to individually process the sequences of immune-related genes' promoters for our "top 10" immune-related modules. The promoter sequences of the genome (33,323 genes) were used as background to compute the fold enrichments of putative *cis*-regulatory elements. We also analyzed the occurrences of all the putative*cis*-regulatory elements (cluster mean), performed 1,000 × bootstrapping to determine the background mean and finally calculated fold enrichment ratios for all six-mer sequences. Three different fold enrichment ratios (≥4, ≥ 3 and≥2) prioritized newly identified *cis*-regulatory elements in "top 10" immune-related modules. To analyze the performance and robustness of our newly developed software, we compared the efficacy of OCCEAN with MEME. We generated a list of experimentally known *cis*-regulatory elements and computed the precision of both OCCEAN and MEME using *cis*-regulatory elements identified in our top 10 immune-related modules. MEME was unable to compute the analyses on top 4 immune-related modules as it can't process over 60,000 characters simultaneously. We computed true positives (nTPs), false positives (nFPs) and precision for MEME as well as OCCEAN enrichment ratios≥4, ≥ 3 and≥2 (Figure 6). We selected an enrichment ratio≥3 as the optimal value for OCCEAN as it yields a higher number of nTPs, significantly fewer nFPs positives and greater precisions compared with enrichment ratios≥4 and≥2. In summary, we demonstrated that OCCEAN boasts higher capacity (can process the entire genome scale; over 30 million characters in a single run) and features higher precision than MEME (Figure 6). Identification of *cis*-regulatory elements among co-expressed genes could provide additional information for prediction of biological function.

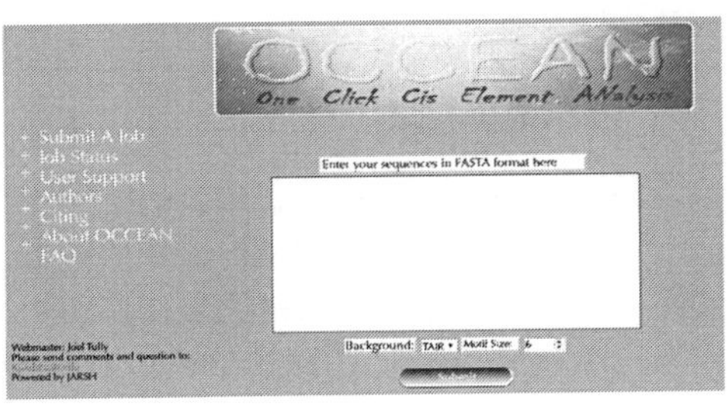

Figure 5: Screenshot of OCCEAN (*O*ne *C*lick *C*is-regulatory *E*lements*AN*alysis). Results can be obtained as an excel sheet and Additional information can be accessed in pop-up windows.

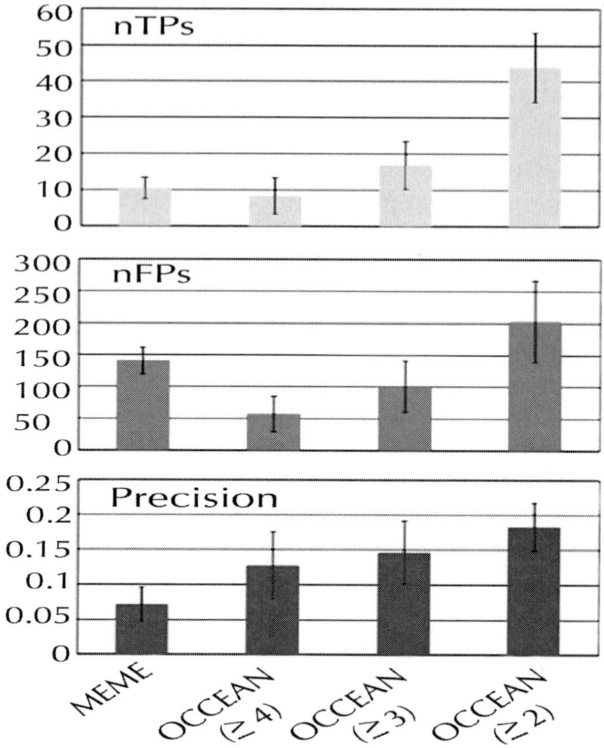

Figure 6: Performance and accuracy of OCCEAN. True positives (nTPs), false positives (nFPs) and precision for MEME as well as OCCEAN with multiple enrichment ratios are shown.

Inferring Biological Relevance of Immune-Related Modules

To fend off potential pathogens, plants employ two major types of immune responses: MAMPs-Triggered Immunity (MTI) and Effector Triggered Immunity (ETI) [4]. ETI is essentially a high amplitude re-booting of the immediate but weaker MTI. ETI results in robust disease resistance responses, including localized host cell death (hypersensitive response; HR) and systemic signaling[4]. In addition, upon pathogen recognition at the MTI and/or ETI levels, the plant cell undergoes extensive transcriptional changes involving complex reiterative signaling networks and cross-talk controlled by phytohormones [66].

This leads to the metabolic reprogramming and production of an array of antimicrobial compounds [27]. Here we highlight two major immune-related functional modules, the largest identified Module 1 (Figure 7) as two most representative gene clusters reflecting the broad array of diverse pathogens and immune stimuli used in the microarray experiments employed in our analyses. Module 1 is the largest module identified by our clustering analyses and comprises 178 nodes, 15,720 edges, clustering coefficient of 0.692 and network density of 0.499. The genes of Module 1 are enriched in three major GO categories: (1) kinases/ribonucleotide binding, (2) immune responses/programmed cell death and (3) transmembrane proteins.

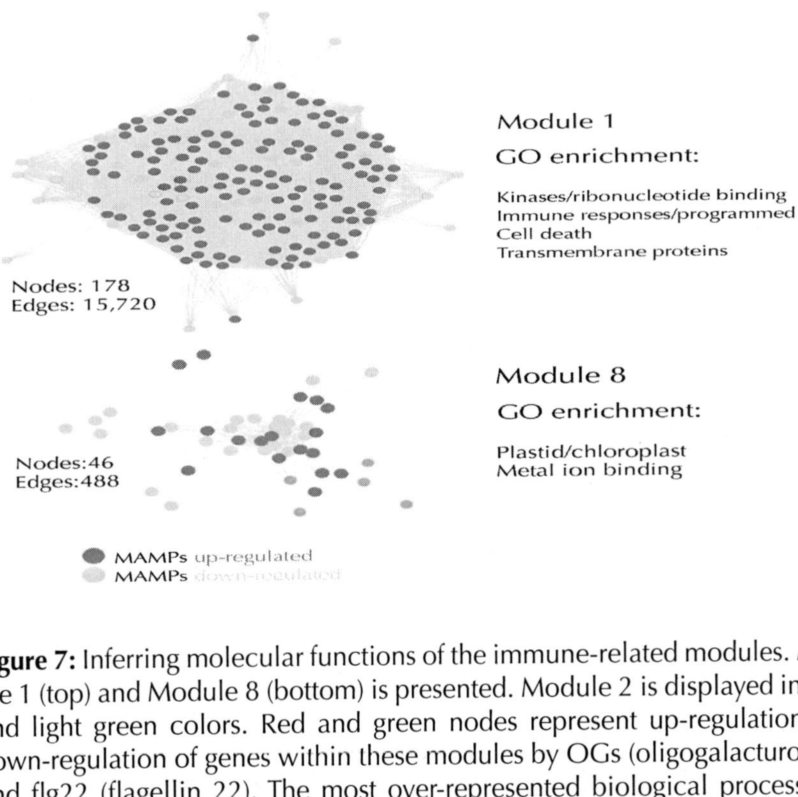

Figure 7: Inferring molecular functions of the immune-related modules. Module 1 (top) and Module 8 (bottom) is presented. Module 2 is displayed in dark and light green colors. Red and green nodes represent up-regulation and down-regulation of genes within these modules by OGs (oligogalacturonide) and flg22 (flagellin 22). The most over-represented biological process GO term was indicated with each module.

Diverse kinases account for the major functional category of Module 1, encompassing 46 out of 178 proteins representing receptor-like kinases (RLKs), mitogen-activated protein kinases (MAPKs), calcium-dependent protein kinases (CDPKs), wall-associated kinases

(WAKs), etc. Typically, RLKs perceive MAMPs such as flg22, elf18, chitin and OGs and trigger MTI [27]. Subsequently, a MAPK cascade is activated and downstream signaling ensues including phytohormonal crosstalk[8,27,44,67]. These events result in the activation of a wide-spectrum of immune responses, such as induced biosynthesis of phytoalexins and other antimicrobial secondary metabolites such as glucosinolates, calcium influx, as well as production and accumulation of reactive oxygen species (ROS) [27,68].

At the second, more powerful layer of the immune response, plants usually deploy NLR receptors (nucleotide-binding domain and leucine-rich repeat receptors; a major class of R proteins). The NLR receptors directly recognize specific effectors or indirectly detect effector activities and trigger ETI[1,2,6,7]. We discovered 8 *NLR* genes in the kinase sub-cluster of Module 1. These include proteins conferring resistance to diverse pathogens such as *Pseudomonas syringae, Albugo candida, Hyaloperonospora arabidopsidis* and various *Fusarium* races.

The second most abundant GO category in Module 1 was immune-related genes (count: 32). This functional sub-cluster contains various defense-related transcription factors such as WRKY15, WRKY33, MYB113 and ANAC061 [9,47], four members of the RING-H2 finger protein family, a heat shock transcription factor HSF A-4a [53], an ethylene-responsive transcription factor ERF105 and, intriguingly, three scarecrow-like GRAS family transcription factor genes, known to be primarily implicated in plant root development [69-71]. Collectively, transcription factor genes account for over 40% of this sub-cluster's gene count. In addition, we identified a small group of genes strongly associated with response to fungal cell wall elicitor chitin, such as two auxin-regulated SAUR genes, one lectin-like, Enhanced Disease Susceptibility 1 (EDS1) and senescence-associated protein SAG102[72-76].

Another enriched GO category in Module 1 was membrane-associated proteins including both plasma membrane- and organellar membrane-resident peptides; not surprisingly, the largest class of these membrane proteins is transporters. In plants, uptake and translocation of nutrients play essential roles in physiological processes including plant growth, nutrition, signal transduction, and development [77-79]. Transport processes are also critical for the reallocation of resources and it was previously shown that defense signals cause realignment of transport activities to redirect resources toward immune responses and protect tissues of high value [80]. Of 31 genes assigned to this sub-cluster, seven were transport proteins (22.6% of this sub-cluster's gene count). The prevailing class was sugar (galactose and sucrose) transporters, in agreement with the fact that this class of proteins constitutes a key component for carbon partitioning at the whole plant level and is involved in both symbiotic and parasitic plant-fungi interactions [81]. Additionally, we identified *CNI1* (*Carbon-Nitrogen Insensitive 1*) that is a key regulator of the Carbon/Nitrogen response for growth phase transition in Arabidopsis seedlings [82], as well as ATP Binding Cassette (ABC) and inorganic phosphate transporters that were previously postulated to be required for organ growth, nutrition, development, and stress responses [83]. Last but not least, we also identified two other well-known immune regulators: *Syntaxin 121/PEN1* and *MLO2* genes, required for resistance against barley powdery mildew, *Blumeria graminis* sp. *hordei* and a fungal pathogen *Colletotrichum higginsianum*[84,85].

Another essential aspect of large-scale transcriptomic analyses is to link co-expression with co-regulation, *i.e.,* to determine presence of common *cis*-regulatory elements among genes of the same module. Several known *cis*-regulatory elements were discovered in the promoters of Module 1 genes. Two of the enriched elements, *TCATGG* and *CATGGA*, overlap with the octadecameric CArG box sequence 5'-*CTTACCTTTCATGGATTA*-3', identified in the *APETALA3*promoter [86]. *APETALA3*, a class B homeotic organ identity gene, was originally discovered as the central regulator of petal development in Arabidopsis flowers, but later also shown to be involved in control of the floral meristem proliferation, regulation of flowering time, and other plant reproductive processes [87-89]. Inhibition of reproductive processes is an expected outcome of an immune stimulation, given the recent report describing a transition from growth to defense following immune stimuli treatments in Arabidopsis [53]. Involvement of a developmental regulator in defense responses is not unprecedented and was previously shown for a MYB-related gene *ASYMMETRIC LEAVES 1* (*AS1*) [90].

Jasmonic Acid (JA) and Ethylene (ET) are two plant hormones antagonistic to SA [91] and down-regulation of major JA/ET signaling proteins will promote SA signaling by suppressing the JA/ET pathway [67]. Consistent with this observation, we identified an enriched element, *ATCTTG* that resembles the binding site for Ethylene-

Insensitive 3 (EIN3) and three EIN3-LIKE (EIL) proteins [92] as well as two hexamers *GTCGTC* and *CGTCGT*, which overlap with the core binding site of *JASE2*motif in two JA biosynthetic genes *OPR1/ OPR2*[93]. Recently, EIN3 and EIL1 were shown to negatively regulate MAMP-triggered immunity via a direct binding and down-regulation of the SA biosynthetic gene *SID2*[94]. Given that OPRs are implicated in JA biosynthesis, we expect a fine-tuned interplay between SA and JA/ET signaling pathways [95]. In addition, several novel *cis*-regulatory elements were discovered in our study, and the identity of the cognate transcription factors and their functional relevance will be subject to future studies. Very recently, it was demonstrated that numerous transcription factors can recognize secondary elements, which in some cases are completely sequence-unrelated to the primary element [96]. In conjunction with this study, our analysis of Module 1 shows that ARR14 (a MYB family transcription factor) not only recognizes the known primary *cis*-regulatory element *AGATA/TCG* but also can specifically bind to a previously unknown *cis*-regulatory element *AGATCT*. Thus, it's likely that transcription factor(s) have extended list of binding specificities beyond the currently known *cis*-regulatory elements. We propose that the above described transcription factors and additional unknown regulatory proteins coordinate the gene regulation in this module.

Our data provide further insights into the transcriptional regulatory mechanisms repressing additional putative negative regulators of plant defense upon treatments with pathogen-mimicking stimuli.

CONCLUSIONS

In this study, we used a systems-level network biology approach to construct genome-wide Arabidopsis immune co-expression network and demonstrated that this network shares properties of a 'real-world network'. Topological properties-based partitioning allowed us to unravel 156 distinct immune-related functional modules. We demonstrated that nodes in the same module are not only highly correlated at the expression level but also densely connected to each other. Detailed analyses of two key immune-related modules provided a systems-level understanding of how plant cells coordinate distinct immune signals to orchestrate fine-tuned and pathogen-specific defense responses. Our novel approach to discover *cis*-regulatory elements using OCCEAN is an effective method of solving the issue of finding novel motifs within a sequence set. OCCEAN has advantages over other programs of the same purpose, such as APPLES (Analysis of Plant Promoter-Linked Elements) and MEME [63,97]. APPLES requires finding organisms of a specified relational distance for comparison, which can burden the user, and MEME has the statistical risk of discarding actual existing motifs. These problems are avoided in our solution whilst continuing to maintain a fair amount of focus on client-side simplicity. In addition, OCCEAN has the capacity to be expanded for analyses of other eukaryotes genomes, such as fly and human. Our systems-level approach to examine *cis*-regulatory elements (the putative transcription factor binding sites) in the promoters of the co-expressed genes made it possible to link co-expression to co-regulation of genes in the same module.

METHODS

Data Download, Selection Criteria for Differentially Expressed (DE) Genes Dataset and Promoter Sequence Acquisition

We utilized available transcriptomic data of transcriptional responses extracted from 271 microarray experiments representing nine major immune-related studies [26-34]. Priority was given to well-referenced studies, employing the Affymetrix ATH1 GeneChip array, encompassing the broadest possible spectrum of plant defense responses upon pathogen infections or pathogen-mimicking stimuli treatment. Lists of probes showing significant up- or down-regulation in each experimental condition were compiled, using criteria for significance employed in the respective original study. This led us to the identification of a list 8,377 genes differentially expressed between all treatments [5]. For each of these genes, we acquired 1000 bp upstream of the transcriptional start site from TAIR Version 10 at http://www.arabidopsis.org [98]. These upstream regions were searched for putative transcription factor binding sites.

Network Construction, Topological Properties, Network Clustering and Gene Ontology (GO)

In the linear model, Pearson Correlation Coefficients (PCC) was measured based on the mutual k-nearest neighbor method of Ruan et al.[18,97] with some modifications. In contrast to Ruan et al. [18,97], the k-list for each gene included k-1 other genes, and at least one (and possibly more) gene sharing the smallest of the k PCCs. For example, for a k of 10, the k-list might include 9 genes in order of decreasing PCC and one or more genes having the next lower PCC value. This was done in consideration of the fact that all genes of equal PCC are equally valid as connected nodes. Random networks were constructed by randomly assigning gene expression values (0 = no change, +1 = up-regulated, -1 = down-regulated) to the genes from the immune transcriptional profiling experiments. The network based on PCC-threshold was inferred by first calculating the PCC for every gene-gene pair in the dataset yielding 8,377 × 8,377 PCC values. These values were tested against 0.9 threshold and only the PCC values that were greater than or equal to 0.9 were selected to indicate that there is an edge between this gene pairs. Other values were discarded. R was used to calculate the non-linear mutual information (MI) [99]. First the MI was calculated using parmigene package were the MI for every gene pair was calculated resulting in 8377 × 8377 MI values [25]. Then these values where used by the ARACNE multiplicative (Algorithm for the Reconstruction of Accurate Cellular Networks), ARACNE additive, CLR (Context Likelihood of Relatedness), and MRNET (Minimum Redundancy NETwork) algorithms, using the same package, to infer a weighted adjacency matrix [21-25]. 80% and 90% of the maximum MI values in each weight adjacency matrix were chosen as the threshold to determine the existence of an edge between each pair of genes. If the value is greater than or equal to the threshold, this indicates an edge between a given pair of genes.

A modified and parallelized Cytoscape version 2.8.4 [37,38] and the Clusterviz version 1.2 (FAG-EC algorithm) [37,38,56,100] and ClusterMaker version 1.10 [100,101] plugins were used to visualize this network, calculate its parameters and network topological properties. Network partitioning was performed using FAG-EC algorithm [56,100] and the Markov Clustering Algorithm [102]. For each gene in the network sub-clusters, a Gene Ontology [57,58] molecular function and biological process was assigned from the Database for Annotation, Visualization and Integrated Discovery (DAVID) v6.7[59].

BLAST-based generation of common *cis*-regulatory elements Sequences obtained from TAIR [98] (1000 bp 5′ of the transcription start site) were incorporated into a local BLAST [103,104] database using the formatdb utility that is included in NCBI's package download of BLAST. Each sequence in the local BLAST database was used in turn as a query sequence to find subsequences shared in the promoter regions of a number of genes. The resulting common subsequences were imported into a relational database. Similar sequences with at least 75% identity and an E-value$\leq 1\times 10^{-4}$ were scanned using a sliding window search in order to extract all contained 6-mers. All 6-mers having no more than four consecutive single nucleotide repeats were deemed putative transcription factor binding sites and retained for further computational analysis. This process has been incorporated into our web-accessible tool, OCCEAN.

Statistical Enrichment of *Cis*-Regulatory Elements Using Modified Version of POBO

Putative transcription factor binding sites results from BLAST were analyzed for statistically significant over representation/under representation using the POBO (a promoter bootstrapping program) tool[64]. POBO is an exceptional tool for statistical sequence significance analysis, but is inhibited by the constraint that it can perform analysis only on one *cis*-regulatory element at a time. As our sequence sets were high in number, manual entry would be infeasible. Therefore, a wrapper program was written to take each individual sequence, create a cluster file consisting of the set of promoter identifiers/sequences from the original experimentally found gene promoter dataset that contain the sequence, and run an instance of POBO with the sequence and cluster file as input. The program then extracted the desired information from the resulting HTML files from all of the individual runs and put them in a single spreadsheet for analysis. Running POBO locally required downloading the source code from http://ekhidna.biocenter.helsinki.fi/poxo/pobo/ and setting up a local MySQL database with all known Arabidopsis gene promoters corresponding to ~33,000 genes as the background data. 1 kb promoter sequence for all ~33,000 genes were obtained from the TAIR10 genome release, and used as background for the POBO analysis. The POBO results were converted to spreadsheet format for further analysis. True positives (nTPs), false positives (nFPs) and precision for MEME as well as OCCEAN were calculated as described in [25].

Development of One-Click Cis-Element Analysis (OCCEAN) Software for Genome-Wide Identification of *Cis*-Regulatory Elements

The bulk of OCCEAN on the server-side was developed in the Java language (version 7) using Apache Tomcat (version 7, as well) on a Linux server. A Java Servlet interface was used to communicate with OCCEAN's web interface. OCCEAN automates and integrates the information obtained from BLAST and POBO analysis. This user friendly software requires promoter sequences in FASTA format. Background sequences of the species under study are imbedded in the software. OCCEAN is time-efficient and will return a link to the file containing the results the BLAST and POBO. OCCEAN's web interface was developed using HTML, CSS, Javascript, and AJAX for in-page update notifications.

AUTHORS' CONTRIBUTIONS

MSM and AEH designed the experiments. JPT performed data mining, designed and developed OCCEAN and computed cross-comparison *cis*-regulatory elements experiments. JPT and RW performed *cis*-regulatory element enrichment experiments. HMRA performed PCC analyses, constructed immune co-expression network, computed network topological properties and performed GO annotation analyses. MSM, AEH and AS supervised HMRA, JPT and RW. MSM and AEH wrote the manuscript and all authors edited the manuscript. All authors read and approved the final manuscript.

ACKNOWLEDGEMENTS

The authors are greatly thankful to Dr. Purushotham Bangalore and Mr. Larry Owen for technical support. The authors thank Drs. Karolina Mukhtar and Michelle Amaral, Ms. Cassandra Garbutt and Ms. Jessica Lopez for critical reading of the manuscript and valuable suggestions.

REFERENCES

1. Mukhtar MS: Engineering NLR immune receptors for broad-spectrum disease resistance. *Trends Plant Sci* 2013, 18(9):469-472.
2. Pajerowska-Mukhtar KM, Emerine DK, Mukhtar MS: Tell me more: roles of NPRs in plant immunity. *Trends Plant Sci* 2013, 18(7):402-411.
3. Wulff BB, Horvath DM, Ward ER: Improving immunity in crops: new tactics in an old game. *Curr Opin Plant Biol* 2011, 14(4):468-476.
4. Jones JD, Dangl JL: The plant immune system. *Nature* 2006, 444(7117):323-329.
5. Mukhtar MS, Carvunis AR, Dreze M, Epple P, Steinbrenner J, Moore J, Tasan M, Galli M, Hao T, Nishimura MT, Pevzner SJ, Donovan SE, Ghamsari L, Santhanam B, Romero V, Poulin MM, Gebreab F, Gutierrez BJ, Tam S, Monachello D, Boxem M, Harbort CJ, McDonald N, Gai L, Chen H, He Y, Consortium EUF, Vandenhaute J, Roth FP, Hill DE, *et al.*: Independently evolved virulence effectors converge onto hubs in a plant immune system network. *Science* 2011, 333(6042):596-601.
6. Nishimura MT, Dangl JL: Arabidopsis and the plant immune system. *Plant J Cell Mole Biol* 2010, 61(6):1053-1066.
7. Mukhtar MS, Nishimura MT, Dangl J: NPR1 in plant defense: it's not over 'til it's turned over. *Cell* 2009, 137(5):804-806.
8. Spoel SH, Dong X: How do plants achieve immunity? Defence without specialized immune cells. *Nat Rev Immunol* 2012, 12(2):89-100.

9. Eulgem T, Somssich IE: Networks of WRKY transcription factors in defense signaling. *Curr Opin Plant Biol* 2007, 10(4):366-371.
10. Guo A, He K, Liu D, Bai S, Gu X, Wei L, Luo J: DATF: a database of Arabidopsis transcription factors. *Bioinformatics* 2005, 21(10):2568-2569.
11. Yilmaz A, Mejia-Guerra MK, Kurz K, Liang X, Welch L, Grotewold E: AGRIS: the arabidopsis gene regulatory information server, an update. *Nucleic Acids Res* 2011, 39(Database issue):D1118-D1122.
12. Arabidopsis Interactome Mapping C: Evidence for network evolution in an Arabidopsis interactome map. *Science* 2011, 333(6042):601-607.
13. Carrera J, Rodrigo G, Jaramillo A, Elena SF: Reverse-engineering the Arabidopsis thaliana transcriptional network under changing environmental conditions. *Genome Biol* 2009, 10(9):R96.
14. Mao L, Van Hemert JL, Dash S, Dickerson JA: Arabidopsis gene co-expression network and its functional modules. *BMC Bioinformatics* 2009, 10:346.
15. Vidal M, Cusick ME, Barabasi AL: Interactome networks and human disease. *Cell* 2011, 144(6):986-998.
16. Ma S, Bohnert HJ: Integration of Arabidopsis thaliana stress-related transcript profiles, promoter structures, and cell-specific expression. *Genome Biol* 2007, 8(4):R49.
17. Heyndrickx KS, Vandepoele K: Systematic identification of functional plant modules through the integration of complementary data sources. *Plant Physiol* 2012, 159(3):884-901.
18. Ruan J, Perez J, Hernandez B, Lei C, Sunter G, Sponsel VM: Systematic identification of functional modules and cis-regulatory elements in Arabidopsis thaliana. *BMC Bioinformatics* 2011, 12(Suppl 12):S2.
19. Song L, Langfelder P, Horvath S: Comparison of co-expression measures: mutual information, correlation, and model based indices. *BMC Bioinformatics* 2012, 13:328.
20. Ruan J, Zhang W: Identifying network communities with a high resolution. *Physical Rev E Stat Nonlinear Soft Matter Physics* 2008, 77(1 Pt 2):016104.

21. Margolin AA, Nemenman I, Basso K, Wiggins C, Stolovitzky G, Dalla Favera R, Califano A:ARACNE: an algorithm for the reconstruction of gene regulatory networks in a mammalian cellular context. *BMC Bioinformatics* 2006, 7(Suppl 1):S7.
22. Zoppoli P, Morganella S, Ceccarelli M: TimeDelay-ARACNE: Reverse engineering of gene networks from time-course data by an information theoretic approach. *BMC Bioinformatics* 2010, 11:154.
23. Olsen C, Meyer PE, Bontempi G: On the impact of entropy estimation on transcriptional regulatory network inference based on mutual information. *EURASIP J Bioinformatics Systems Biol* 2009, 2009(1):308959.
24. Meyer PE, Kontos K, Lafitte F, Bontempi G: Information-theoretic inference of large transcriptional regulatory networks. *EURASIP J Bioinformatics Systems Biol* 2007, 2007:79879.
25. Sales G, Romualdi C: Parmigene–a parallel R package for mutual information estimation and gene network reconstruction. *Bioinformatics* 2011, 27(13):1876-1877.
26. Chandran D, Inada N, Hather G, Kleindt CK, Wildermuth MC: Laser microdissection of Arabidopsis cells at the powdery mildew infection site reveals site-specific processes and regulators. *Proc Nat Acad Sci USA* 2010, 107(1):460-465.
27. Denoux C, Galletti R, Mammarella N, Gopalan S, Werck D, De Lorenzo G, Ferrari S, Ausubel FM, Dewdney J: Activation of defense response pathways by OGs and Flg22 elicitors in Arabidopsis seedlings. *Mole Plant* 2008, 1(3):423-445.
28. Eulgem T, Weigman VJ, Chang HS, McDowell JM, Holub EB, Glazebrook J, Zhu T, Dangl JL:Gene expression signatures from three genetically separable resistance gene signaling pathways for downy mildew resistance. *Plant Physiol* 2004, 135(2):1129-1144
29. Goda H, Sasaki E, Akiyama K, Maruyama-Nakashita A, Nakabayashi K, Li W, Ogawa M, Yamauchi Y, Preston J, Aoki K, Kiba T, Takatsuto S, Fujioka S, Asami T, Nakano T, Kato H, Mizuno T, Sakakibara H, Yamaguchi S, Nambara E, Kamiya Y, Takahashi H, Hirai MY, Sakurai T, Shinozaki K, Saito K, Yoshida S, Shimada Y: The AtGenExpress hormone and chemical treatment data set:

experimental design, data evaluation, model data analysis and data access. *Plant J Cell Mole Biol* 2008, 55(3):526-542.

30. Ramonell K, Berrocal-Lobo M, Koh S, Wan J, Edwards H, Stacey G, Somerville S: Loss-of-function mutations in chitin responsive genes show increased susceptibility to the powdery mildew pathogen Erysiphe cichoracearum. *Plant Physiol* 2005, 138(2):1027-1036.
31. Thilmony R, Underwood W, He SY: Genome-wide transcriptional analysis of the Arabidopsis thaliana interaction with the plant pathogen Pseudomonas syringae pv. tomato DC3000 and the human pathogen Escherichia coli O157:H7. *Plant J Cell Mole Biol* 2006, 46(1):34-53.
32. Truman W, de Zabala MT, Grant M: Type III effectors orchestrate a complex interplay between transcriptional networks to modify basal defence responses during pathogenesis and resistance. *Plant J Cell Mole Biol* 2006, 46(1):14-33.
33. Wang D, Amornsiripanitch N, Dong X: A genomic approach to identify regulatory nodes in the transcriptional network of systemic acquired resistance in plants. *PLoS Pathogens* 2006, 2(11):e123.
34. Zipfel C, Kunze G, Chinchilla D, Caniard A, Jones JD, Boller T, Felix G: Perception of the bacterial PAMP EF-Tu by the receptor EFR restricts Agrobacterium-mediated transformation. *Cell* 2006, 125(4):749-760.
35. Altay G, Emmert-Streib F: Inferring the conservative causal core of gene regulatory networks. *BMC Systems Biol* 2010, 4:132.
36. Luo F, Yang Y, Zhong J, Gao H, Khan L, Thompson DK, Zhou J: Constructing gene co-expression networks and predicting functions of unknown genes by random matrix theory. *BMC Bioinformatics* 2007, 8:299.
37. Cline MS, Smoot M, Cerami E, Kuchinsky A, Landys N, Workman C, Christmas R, Avila-Campilo I, Creech M, Gross B, Hanspers K, Isserlin R, Kelley R, Killcoyne S, Lotia S, Maere S, Morris J, Ono K, Pavlovic V, Pico AR, Vailaya A, Wang PL, Adler A, Conklin BR, Hood L, Kuiper M, Sander C, Schmulevich I, Schwikowski B, Warner GJ, *et al.*: Integration of biological networks and gene expression data using Cytoscape. *Nat. Protoc* 2007, 2(10):2366-2382.

38. Smoot ME, Ono K, Ruscheinski J, Wang PL, Ideker T: Cytoscape 2.8: new features for data integration and network visualization. *Bioinformatics* 2011, 27(3):431-432.

39. Seebacher J, Gavin AC: SnapShot: Protein-protein interaction networks. *Cell* 2011, 144(6):1000. 1000 e1001.

40. Batada NN, Reguly T, Breitkreutz A, Boucher L, Breitkreutz BJ, Hurst LD, Tyers M: Stratus not altocumulus: a new view of the yeast protein interaction network. *PLoS Biol* 2006, 4(10):e317.

41. Dangl JL, Jones JD: Plant pathogens and integrated defence responses to infection. *Nature* 2001, 411(6839):826-833.

42. van der Biezen EA, Freddie CT, Kahn K, Parker JE, Jones JD: Arabidopsis RPP4 is a member of the RPP5 multigene family of TIR-NB-LRR genes and confers downy mildew resistance through multiple signalling components. *Plant J Cell Mole Biol* 2002, 29(4):439-451.

43. Knoth C, Ringler J, Dangl JL, Eulgem T: Arabidopsis WRKY70 is required for full RPP4-mediated disease resistance and basal defense against Hyaloperonospora parasitica. *MPMI* 2007, 20(2):120-128.

44. Rasmussen MW, Roux M, Petersen M, Mundy J: MAP kinase cascades in arabidopsis innate immunity. *Front Plant Sci* 2012, 3:169.

45. Li J, Besseau S, Toronen P, Sipari N, Kollist H, Holm L, Palva ET: Defense-related transcription factors WRKY70 and WRKY54 modulate osmotic stress tolerance by regulating stomatal aperture in Arabidopsis. *New Phytol* 2013, 200(2):457-472.

46. Besseau S, Li J, Palva ET: WRKY54 and WRKY70 co-operate as negative regulators of leaf senescence in Arabidopsis thaliana. *J Exp Bot* 2012, 63(7):2667-2679.

47. Jensen MK, Hagedorn PH, de Torres-Zabala M, Grant MR, Rung JH, Collinge DB, Lyngkjaer MF:Transcriptional regulation by an NAC (NAM-ATAF1,2-CUC2) transcription factor attenuates ABA signalling for efficient basal defence towards Blumeria graminis f. sp. hordei in Arabidopsis. *Plant J Cell Mole Biol* 2008, 56(6):867-880.

48. Bassham DC, Brandizzi F, Otegui MS, Sanderfoot AA: The secretory system of Arabidopsis. *Arabidopsis Book/Am Soc Plant Biol* 2008, 6:e0116.

49. Mishiba K, Nagashima Y, Suzuki E, Hayashi N, Ogata Y, Shimada Y, Koizumi N: Defects in IRE1 enhance cell death and fail to degrade mRNAs encoding secretory pathway proteins in the Arabidopsis unfolded protein response. *Proc Nat Acad Sci USA* 2013, 110(14):5713-5718.

50. Tierens KF, Thomma BP, Brouwer M, Schmidt J, Kistner K, Porzel A, Mauch-Mani B, Cammue BP, Broekaert WF: Study of the role of antimicrobial glucosinolate-derived isothiocyanates in resistance of Arabidopsis to microbial pathogens. *Plant Physiol* 2001, 125(4):1688-1699.

51. Damon C, Dmitrieva J, Muhovski Y, Francis F, Lins L, Ledoux Q, Luwaert W, Marko IE, Mauro S, Ongena M, Thonart P, Veys P, Portetelle D, Twizere JC, Vandenbol M: Interaction network of antimicrobial peptides of Arabidopsis thaliana, based on high-throughput yeast two-hybrid screening. *Plant Physiol Biochem PPB/Societe francaise de physiologie vegetale* 2012, 58:245-252.

52. Boatwright JL, Pajerowska-Mukhtar K: Salicylic acid: an old hormone up to new tricks. *Mole Plant Pathol* 2013, 14(6):623-634.

53. Pajerowska-Mukhtar KM, Wang W, Tada Y, Oka N, Tucker CL, Fonseca JP, Dong X: The HSF-like transcription factor TBF1 is a major molecular switch for plant growth-to-defense transition. *CB* 2012, 22(2):103-112.

54. Goritschnig S, Zhang Y, Li X: The ubiquitin pathway is required for innate immunity in Arabidopsis. *Plant J Cell Mole Biol* 2007, 49(3):540-551.

55. Trujillo M, Ichimura K, Casais C, Shirasu K: Negative regulation of PAMP-triggered immunity by an E3 ubiquitin ligase triplet in Arabidopsis. *CB* 2008, 18(18):1396-1401.

56. Wang J, Li M, Deng Y, Pan Y: Recent advances in clustering methods for protein interaction networks. *BMC Genomics* 2010, 11(Suppl 3):S10.

57. Ashburner M, Ball CA, Blake JA, Botstein D, Butler H, Cherry JM, Davis AP, Dolinski K, Dwight SS, Eppig JT, Harris MA, Hill

DP, Issel-Tarver L, Kasarskis A, Lewis S, Matese JC, Richardson JE, Ringwald M, Rubin GM, Sherlock G: Gene ontology: tool for the unification of biology. The Gene Ontology Consortium. *Nat Genet* 2000, 25(1):25-29.

58. Gaudet P, Livstone MS, Lewis SE, Thomas PD: Phylogenetic-based propagation of functional annotations within the Gene Ontology consortium. *Brief Bioinform*. 2011, 12(5):449-462.

59. Jiao X, Sherman BT, da Huang W, Stephens R, Baseler MW, Lane HC, Lempicki RA: DAVID-WS: a stateful web service to facilitate gene/protein list analysis. *Bioinformatics* 2012, 28(13):1805-1806.

60. Higo K, Ugawa Y, Iwamoto M, Korenaga T: Plant cis-acting regulatory DNA elements (PLACE) database: 1999. *Nucleic Acids Res* 1999, 27(1):297-300.

61. O'Connor TR, Dyreson C, Wyrick JJ: Athena: a resource for rapid visualization and systematic analysis of Arabidopsis promoter sequences. *Bioinformatics* 2005, 21(24):4411-4413.

62. Palaniswamy SK, James S, Sun H, Lamb RS, Davuluri RV, Grotewold E: AGRIS and AtRegNet. a platform to link cis-regulatory elements and transcription factors into regulatory networks. *Plant Physiol* 2006, 140(3):818-829.

63. Bailey TL, Boden M, Buske FA, Frith M, Grant CE, Clementi L, Ren J, Li WW, Noble WS: MEME SUITE: tools for motif discovery and searching. *Nucleic Acids Res* 2009, 37(Web Server issue):W202-W208.

64. Kankainen M, Holm L: POBO, transcription factor binding site verification with bootstrapping. *Nucleic Acids Res* 2004, 32(Web Server issue):W222-W229.

65. Bulow L, Brill Y, Hehl R: AthaMap-assisted transcription factor target gene identification in Arabidopsis thaliana. *Database J Biol Databases Curation* 2010, 2010:baq034.

66. Newman MA, Sundelin T, Nielsen JT, Erbs G: MAMP (microbe-associated molecular pattern) triggered immunity in plants. *Front Plant Sci* 2013, 4:139.

67. Spoel SH, Dong X: Making sense of hormone crosstalk during plant immune responses. *Cell Host Microbe* 2008, 3(6):348-351.

68. Mersmann S, Bourdais G, Rietz S, Robatzek S: Ethylene signaling regulates accumulation of the FLS2 receptor and is required for the oxidative burst contributing to plant immunity. *Plant Physiol* 2010, 154(1):391-400.

69. Meng X, Xu J, He Y, Yang KY, Mordorski B, Liu Y, Zhang S: Phosphorylation of an ERF transcription factor by Arabidopsis MPK3/MPK6 regulates plant defense gene induction and fungal resistance. *Plant Cell* 2013, 25(3):1126-1142.

70. Nakano T, Suzuki K, Fujimura T, Shinshi H: Genome-wide analysis of the ERF gene family in Arabidopsis and rice. *Plant Physiol* 2006, 140(2):411-432.

71. Gobbato E, Marsh JF, Vernie T, Wang E, Maillet F, Kim J, Miller JB, Sun J, Bano SA, Ratet P, Mysore KS, Dénarié J, Schultze M, Oldroyd GE: A GRAS-type transcription factor with a specific function in mycorrhizal signaling. *CB* 2012, 22(23):2236-2241.

72. Schon M, Toller A, Diezel C, Roth C, Westphal L, Wiermer M, Somssich IE: Analyses of wrky18 wrky40 plants reveal critical roles of SA/EDS1 signaling and indole-glucosinolate biosynthesis for Golovinomyces orontii resistance and a loss-of resistance towards Pseudomonas syringae pv. tomato AvrRPS4. *MPMI* 2013, 26(7):758-767.

73. Wiermer M, Feys BJ, Parker JE: Plant immunity: the EDS1 regulatory node. *Curr Opin Plant Biol* 2005, 8(4):383-389.

74. Anai T, Kono N, Kosemura S, Yamamura S, Hasegawa K: Isolation and characterization of an auxin-inducible SAUR gene from radish seedlings. *DNA Seq J DNA Sequencing Mapping* 1998, 9(5–6):329-333.

75. Leyser HM, Pickett FB, Dharmasiri S, Estelle M: Mutations in the AXR3 gene of Arabidopsis result in altered auxin response including ectopic expression from the SAUR-AC1 promoter. *Plant J Cell Mole Biol* 1996, 10(3):403-413.

76. Li Y, Liu ZB, Shi X, Hagen G, Guilfoyle TJ: An auxin-inducible element in soybean SAUR promoters. *Plant Physiol* 1994, 106(1):37-43.

77. Tommasini R, Vogt E, Fromenteau M, Hortensteiner S, Matile P, Amrhein N, Martinoia E: An ABC-transporter of Arabidopsis thaliana has both glutathione-conjugate and chlorophyll catabolite transport activity. *Plant J Cell Mole Biol* 1998, 13(6):773-780.

78. Guo WJ, Nagy R, Chen HY, Pfrunder S, Yu YC, Santelia D, Frommer WB, Martinoia E:SWEET17, a facilitative transporter, mediates fructose transport across the tonoplast of Arabidopsis roots and leaves. *Plant Physiol* 2014, 164(2):777-789.

79. Shkolnik-Inbar D, Adler G, Bar-Zvi D: ABI4 downregulates expression of the sodium transporter HKT1;1 in Arabidopsis roots and affects salt tolerance. *Plant J Cell Mole Biol* 2013, 73(6):993-1005.

80. Nour-Eldin HH, Andersen TG, Burow M, Madsen SR, Jorgensen ME, Olsen CE, Dreyer I, Hedrich R, Geiger D, Halkier BA: NRT/PTR transporters are essential for translocation of glucosinolate defence compounds to seeds. *Nature* 2012, 488(7412):531-534.

81. Doidy J, Grace E, Kuhn C, Simon-Plas F, Casieri L, Wipf D: Sugar transporters in plants and in their interactions with fungi. *Trends Plant Sci* 2012, 17(7):413-422.

82. Sato T, Maekawa S, Yasuda S, Sonoda Y, Katoh E, Ichikawa T, Nakazawa M, Seki M, Shinozaki K, Matsui M, Goto DB, Ikeda A, Yamaguchi J: CNI1/ATL31, a RING-type ubiquitin ligase that functions in the carbon/nitrogen response for growth phase transition in Arabidopsis seedlings. *Plant J Cell Mole Biol* 2009, 60(5):852-864.

83. Kang J, Park J, Choi H, Burla B, Kretzschmar T, Lee Y, Martinoia E: Plant ABC Transporters. *Arabidopsis book / Am Soc Plant Biol* 2011, 9:e0153.

84. Consonni C, Bednarek P, Humphry M, Francocci F, Ferrari S, Harzen A, Ver Loren van Themaat E, Panstruga R: Tryptophan-derived metabolites are required for antifungal defense in the Arabidopsis mlo2 mutant. *Plant Physiol* 2010, 152(3):1544-1561.

85. Lorek J, Griebel T, Jones AM, Kuhn H, Panstruga R: The role of Arabidopsis heterotrimeric G-protein subunits in MLO2 function and MAMP-triggered immunity. *MPMI* 2013, 26(9):991-1003

86. Tilly JJ, Allen DW, Jack T: The CArG boxes in the promoter of the Arabidopsis floral organ identity gene APETALA3 mediate diverse regulatory effects. *Development* 1998, 125(9):1647-1657.

87. Mara CD, Huang T, Irish VF: The Arabidopsis floral homeotic proteins APETALA3 and PISTILLATA negatively regulate the BANQUO genes implicated in light signaling. *Plant Cell* 2010, 22(3):690-702.

88. Krizek BA, Meyerowitz EM: The Arabidopsis homeotic genes APETALA3 and PISTILLATA are sufficient to provide the B class organ identity function. *Development* 1996, 122(1):11-22

89. Purugganan MD, Suddith JI: Molecular population genetics of floral homeotic loci. Departures from the equilibrium-neutral model at the APETALA3 and PISTILLATA genes of Arabidopsis thaliana. *Genetics* 1999, 151(2):839-848.

90. Nurmberg PL, Knox KA, Yun BW, Morris PC, Shafiei R, Hudson A, Loake GJ: The developmental selector AS1 is an evolutionarily conserved regulator of the plant immune response. *Proc Nat Acad Sci USA* 2007, 104(47):18795-18800.

91. Denance N, Sanchez-Vallet A, Goffner D, Molina A: Disease resistance or growth: the role of plant hormones in balancing immune responses and fitness costs. *Front Plant Sci* 2013, 4:155.

92. Solano R, Stepanova A, Chao Q, Ecker JR: Nuclear events in ethylene signaling: a transcriptional cascade mediated by ETHYLENE-INSENSITIVE3 and ETHYLENE-RESPONSE-FACTOR1. *Genes Dev* 1998, 12(23):3703-3714.

93. He Y, Gan S: Identical promoter elements are involved in regulation of the OPR1 gene by senescence and jasmonic acid in Arabidopsis. *Plant Mole Biol* 2001, 47(5):595-605.

94. Chen H, Xue L, Chintamanani S, Germain H, Lin H, Cui H, Cai R, Zuo J, Tang X, Li X, Guo H, Zhou JM: Ethylene insensitive3 and ethylene insensitive3-like1 repress salicylic acid induction deficient2 expression to negatively regulate plant innate immunity in Arabidopsis. *Plant Cell* 2009, 21(8):2527-2540.

95. Chehab EW, Kim S, Savchenko T, Kliebenstein D, Dehesh K, Braam J: Intronic T-DNA insertion renders Arabidopsis opr3 a conditional jasmonic acid-producing mutant. *Plant Physiol* 2011, 156(2):770-778.

96. Franco-Zorrilla JM, Lopez-Vidriero I, Carrasco JL, Godoy M, Vera P, Solano R: DNA-binding specificities of plant transcription factors and their potential to define target genes. *Proc Nat Acad Sci USA* 2014, 111(6):2367-2372.

97. Baxter L, Jironkin A, Hickman R, Moore J, Barrington C, Krusche P, Dyer NP, Buchanan-Wollaston V, Tiskin A, Beynon J, Denby K, Ott S: Conserved noncoding sequences highlight shared

components of regulatory networks in dicotyledonous plants. *Plant Cell* 2012, 24(10):3949-3965.

98. Lamesch P, Berardini TZ, Li D, Swarbreck D, Wilks C, Sasidharan R, Muller R, Dreher K, Alexander DL, Garcia-Hernandez M, Karthikeyan AS, Lee CH, Nelson WD, Ploetz L, Singh S, Wensel A, Huala E: The Arabidopsis Information Resource (TAIR): improved gene annotation and new tools. *Nucleic Acids Res* 2012, 40(Database issue):D1202-D1210.
99. Team RC: R: A Language and Environment for Statistical Computing. In *R Foundation for Statistical Computing*. Vienna, Austria; 2013. http://www.R-project.org
100. Saito R, Smoot ME, Ono K, Ruscheinski J, Wang PL, Lotia S, Pico AR, Bader GD, Ideker T: A travel guide to cytoscape plugins. *Nat Met* 2012, 9(11):1069-1076
101. Morris JH, Apeltsin L, Newman AM, Baumbach J, Wittkop T, Su G, Bader GD, Ferrin TE:clusterMaker: a multi-algorithm clustering plugin for Cytoscape. *BMC Bioinformatics* 2011, 12:436.
102. Enright AJ, Van Dongen S, Ouzounis CA: An efficient algorithm for large-scale detection of protein families. *Nucleic Acids Res* 2002, 30(7):1575-1584.
103. Altschul SF, Gish W, Miller W, Myers EW, Lipman DJ: Basic local alignment search tool. *J Mole Biol* 1990, 215(3):403-410.
104. Mount DW: Using the Basic Local Alignment Search Tool (BLAST). *CSH Protoc* 2007, 2007:pdb top17.

Chapter 2

Functional Genomics of a Generalist Parasitic Plant: Laser Microdissection of Host-Parasite Interface Reveals Host-Specific Patterns of Parasite Gene Expression

Loren A Honaas[1], Eric K Wafula[2], Zhenzhen Yang[1], Joshua P Der1,2, Norman J Wickett[1,2,8],Naomi S Altman[3], Christopher G Taylor[4], John I Yoder[5], Michael P Timko[6], James H Westwood[7], and Claude W dePamphilis[1,2]

[1]Intercollege Graduate Program in Plant Biology, Huck Institutes of the Life Sciences, The Pennsylvania State University, University Park, Pennsylvania 16802, USA

[2]Department of Biology and Institute of Molecular Evolutionary Genetics, The Pennsylvania State University, University Park, Pennsylvania 16802, USA

[3]Department of Statistics and Huck Institutes of the Life Sciences, The Pennsylvania State University, University Park, Pennsylvania, 16802, USA

[4]Department of Plant Pathology, The Ohio State University, Ohio Agricultural Research and Development Center, Wooster, OH, 44691, USA

[5]Department of Plant Sciences, University of California, Davis, Davis, California, 95616, USA

[6]Department of Biology, University of Virginia, Charlottesville, VA, 22904, USA

[7]Department of Plant Pathology, Physiology and Weed Science, Virginia Polytechnic Institute and State University, Blacksburg, VA, 24061, USA

[8]Present address: Chicago Botanic Garden, Glencoe, IL, 60022, USA

ABSTRACT

Background

Orobanchaceae is the only plant family with members representing the full range of parasitic lifestyles plus a free-living lineage sister to all parasitic lineages, *Lindenbergia*. A generalist member of this family, and an important parasitic plant model, *Triphysaria versicolor* regularly feeds upon a wide range of host plants. Here, we compare *de novo* assembled transcriptomes generated from laser micro-dissected tissues at the host-parasite interface to uncover details of the largely uncharacterized interaction between parasitic plants and their hosts.

Results

The interaction of *Triphysaria* with the distantly related hosts *Zea mays* and *Medicago truncatula*reveals dramatic host-specific gene expression patterns. Relative to above ground tissues, gene families are disproportionally represented at the interface including enrichment for transcription factors and genes of unknown function. Quantitative Real-Time PCR of a *T. versicolor* β-expansin shows strong differential

(120x) upregulation in response to the monocot host *Z. mays*; a result that is concordant with our read count estimates. Pathogenesis-related proteins, other cell wall modifying enzymes, and orthologs of genes with unknown function (annotated as such in sequenced plant genomes) are among the parasite genes highly expressed by *T. versicolor* at the parasite-host interface.

Conclusions

Laser capture microdissection makes it possible to sample the small region of cells at the epicenter of parasite host interactions. The results of our analysis suggest that *T. versicolor*'s generalist strategy involves a reliance on overlapping but distinct gene sets, depending upon the host plant it is parasitizing. The massive upregulation of a *T. versicolor* β-expansin is suggestive of a mechanism for parasite success on grass hosts. In this preliminary study of the interface transcriptomes, we have shown that *T. versicolor*, and the Orobanchaceae in general, provide excellent opportunities for the characterization of plant genes with unknown functions.

BACKGROUND

Approximately 1% of all extant angiosperm species are parasitic, deriving all or part of the water and nutrients from host plant species using specialized feeding structures known as haustoria. Among families containing parasites, only the Orobanchaceae contain species representing the full spectrum of parasitism from potentially free-living facultative forms to non-photosynthetic, obligate parasites [1]. *Lindenbergia* is a non-parasitic lineage of Orobanchaceae sister to all parasitic species [2]; therefore the family represents an ideal comparative framework to study the evolution of parasitism. Parasitic Orobanchaceae growing in Africa and the Mediterranean include the devastating agricultural pests witchweed (*Striga*) and broomrape (*Orobanche* and*Phelipanche*), respectively. The *Striga* infestation covers 123.5 million acres resulting in annual yield losses greater than US$7 billion [3,4]. Broomrapes threaten nearly 40 million acres, though yield losses are difficult to assess due to the frequent abandonment of infested fields and unreliable data on yield loss [5]. *Striga* is one of the primary biotic constraints to agriculture in Sub-Saharan Africa and the affected areas are increasing in size [6]. The weedy, parasitic Orobanchaceae also threaten parts of Asia, Europe, and North America [7].

Motivated by the agronomic threat presented by some parasitic Orobanchaceae, *Triphysaria versicolor* has been developed as a model parasitic plant for the family. As a transformable [8] and tractable facultative generalist parasite, *T. versicolor* represents an excellent species to investigate the evolution of parasitism, haustorium development, plant-plant communication, host-parasite interactions, and many other facets of parasite biology [9,10]. To discover processes important in parasitic plant biology, we focused our analysis on the unifying anatomical feature of parasitic plants, the haustorium. This modified root structure is adapted to enable feeding on the host and is unique to parasitic plants, thus it is a focal point for interactions between the parasite and host [11].

Heide-Jorgensen and Kuijt [12,13] showed that the haustorium of *T. versicolor* contains many specialized cells including haustorial hairs, a xylem bridge between the host and parasite, and transfer-like cells adjacent to vessel elements at the host-parasite interface. Although histological evidence for xylem connectivity between the haustorium of *T. versicolor* and its host is well documented [12,13], there is no evidence for phloem connectivity. However, there is evidence that phloem-mobile virus particles move between host and parasite in the holoparasite *Phelipanche*(syn. *Orobanche*) *ramosa*[14] and phloem continuity has been observed in *Orobanche crenata*[15]. The mechanisms of transport between host phloem and parasite phloem likely vary in different parasites from direct phloem connections [15] to transport via apoplastic pathways. Bi-directional movement of small RNAs between host and parasite has been documented in *T.versicolor* attacking transgenic lettuce [16]. The anatomy of the haustorial interface cells and empirical evidence for bi-directional transport point to the host-parasite interface as an epicenter of host-parasite dialogue.

Intimate symbioses tend towards specialization (e.g. parasitism) [17]. A true generalist strategy, where a parasite routinely feeds on many distantly related host species, is relatively uncommon in parasitic organisms [18]. At face value, this is surprising, because a broad host range provides more feeding opportunities. Seedlings of most parasitic plants, for example, must contact and parasitize a suitable host plant soon after germination [10], and access to a wider range of potential host plants should *increase* the likelihood of survival, regardless of the specific plants growing nearby [19]. Although less common than host plant specialists, many parasitic plant families do contain generalists, including some or all parasitic members of Orobanchaceae, Lauraceae, Convolvulaceae, Krameriaceae, and most of the 18 families of Santalales (sandalwoods, mistletoes and their relatives [20]).

If mutations that increase specialized feeding strategies increase in frequency when specific host resources are predictable [21], then traits associated with maintenance of generalist abilities are likely to decrease in frequency. If a generalist strategy involves the evolution of a general-purpose suite of genes that are necessary and sufficient to successfully parasitize a wide range of hosts, then such a trend could lead to a long-term stable generalist strategy. Alternatively, if generalists maintain distinct sets of genes specific to different hosts, then the long-term maintenance of gene sets for attacking different hosts may be unlikely unless there is frequent reinforcement by a diverse range of hosts.

Triphysaria (Orobanchaceae) is a generalist parasite that feeds on a highly diverse collection of angiosperms in nature, including at least 30 species in 17 families of monocot and eudicot host plants [22]. We reasoned that sequencing transcriptomes from the haustorium of *T. versicolor*grown on distantly related hosts would maximize the potential to identify both shared and host-specific patterns of gene expression. The transcriptome datasets of *T. versicolor* provide a unique opportunity to leverage newly established genomic resources of the Parasitic Plant Genome Project (PPGP, [23]) with well-developed functional protocols including parasite-host co-culture [9,24], haustorium induction assays [25], and parasite transformation [8,16,26]. By characterizing the molecular signature of host-parasite interactions, we stand to gain insight into the processes underway in a generalist parasite that facilitate a broad host range and learn about the molecular mechanisms that can facilitate the generalist parasite strategy.

Two substantial hurdles emerge when characterizing the transcriptomes of *T. versicolor* haustoria. The first is that gene expression profiles of specialized cells in the haustorium become diluted when harvesting even the tiny haustorium (1–2 mm diameter) of *T. versicolor.*

The excellent histology and electron microscopy work by Heide-Jorgensen and Kuijt [12,13] revealed cells residing at the host parasite interface that had transfer cell-like morphology. The anatomy of these specialized cells includes dense cytoplasm, numerous small vacuoles, a highly invaginated cell membrane, and a labyrinthine cell wall (for a review see [27]). We hypothesized that the small collection of interface cells, including those with transfer-cell like morphology, facilitate the elusive molecular interaction between host and parasite, making them excellent candidates for transcriptome analysis. The second hurdle is that discovery of genes and subsequent gene expression analysis on a genome-wide scale is difficult without a sequenced and well-annotated genome, which is currently lacking for *T. versicolor*. Next Generation Sequencing (NGS) technologies have emerged as powerful tools for exploring new genomes because the cost per base is substantially lower than traditional dye-terminator or even pyro-sequencing (454) methods[28]. In the wake of the NGS revolution several tools for data analysis (for a review see [29]), including high performance *de novo* transcriptome assemblers like Trinity [30], have emerged to facilitate transcriptome analysis in uncharacterized model systems.

To overcome the limitations of reference independent transcriptome analysis of small numbers of difficult to harvest cells, we developed methods to sample parasite-host interface cells from *T. versicolor* grown on the distantly related and sequenced model hosts *Zea mays* (B73) (monocot) and *Medicago truncatula* (A17) (eudicot) via Laser Pressure Catapult Microdissection (LPCM). We extracted and then amplified exceedingly small RNA samples via T7-based linear amplification and then deeply sequenced each of the amplified parasite-host interface transcriptomes. We assembled millions of paired-end Illumina reads *de novo,* annotated each assembly and then estimated levels of gene expression via read mapping to the *de novo* assembled transcriptome. Using this approach, we identified genes that were part of a host-specific response as well as those that are part of a shared response of *T. versicolor* to the different hosts. We also verified the host-specific differential expression pattern of two *Triphysaria* expansin genes. Expansins are among the few genes known to be differentially regulated in haustoria [31,32]. Analysis of expansin genes allowed us to verify the differential gene expression pattern present in the interface sequence data, and demonstrate the first evidence that a β-expansin is highly upregulated in *T. versicolor*when grown on the *Z. mays* host. Our results suggest that the maintenance of a generalist feeding strategy in *Triphysaria* involves both generalized and specialized gene responses that help us understand *Triphysaria's* generalist feeding abilities.

RESULTS

Parasite Host Co-Culture and Microdissection of the *T. Versicolor*haustorium

T. versicolor and hosts were germinated and grown axenically in separate culture plates. To begin co-culture, hosts were transferred to fresh plates and *T. versicolor* were added and placed in close proximity (~1 mm) to host roots. The attachment rate of *T. versicolor* to host roots was ~90% for*M. truncatula* and ~50% for *Z. mays*. This difference was likely due to the more rapid growth rate of *Z. mays* (compared to *M. truncatula*) coupled with the confined dimensions of the co-culture petri dish rather than differential parasite-host compatibility. Where host roots remained more or less stationary on the agar growth medium during early phases of co-culture, the attachment rates of *T. versicolor* were high (>90%) and equivalent between *Z. mays* and *M. truncatula*.

The first step in sample preparation for LPCM was isolation and cryosectioning of haustoria formed on each host. The optimum section thickness was determined empirically by micro-dissecting samples from sections cut at 1 μm section thickness intervals from 18-30 μm. For *T. versicolor*haustoria we determined that 25 μm cryosections were optimal to allow efficient tissue release from the adhesive coated StarFrost™ LPCM slides coupled with maximum tissue harvest volume. Parasite and host cells that were in contact with each other at the interface were difficult to separate, so to ensure capture of the entire parasite interface cell population we intentionally included a minimal amount of host tissue, knowing that host transcripts could be identified and removed informatically. Figure 1 shows a typical haustorium cross section before (A) and after (B) LPCM.

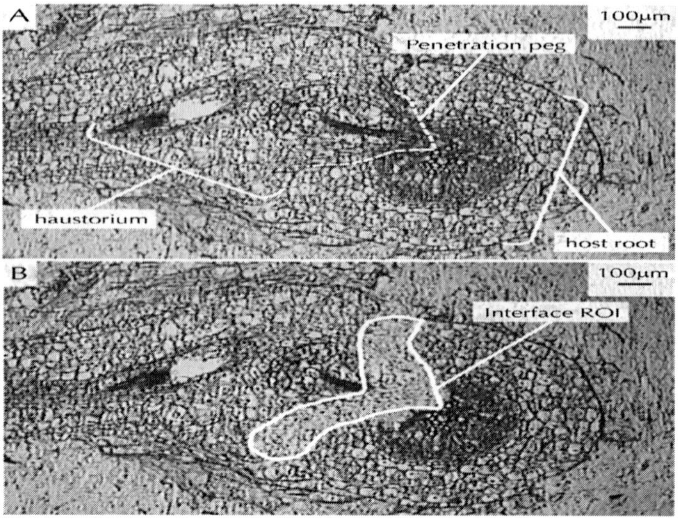

Figure 1: Laser Microdissected Haustorium. LPCM allows highly tissue- and cell-specific harvest after histological identification of tissues or cells of interest. A) Representative 25 μm cross-section of *T. versicolor*haustorium on the host *M. truncatula* approximately 9 days post infestation, and prior to LPCM. The mature haustorium contains the xylem bridge that connects the parasite and host vasculature and is visible in the penetration peg. B) The same section after LCPM shows the cleared interface tissue from the user-defined region of interest (ROI). The flakes of tissue are catapulted by a photonic cloud resulting from pulses of laser light focused between the tissue and glass slide. Multiple pulses of laser light raster across the ROI causing tissue in the selected region to be catapulted and then captured in the adhesive coated cap of a 0.5 mL tube held by a robotic arm in very close proximity (< 0.5 mm) to the upper surface of the section affixed to the slide.

To generate representative interface-cell samples we pooled ~110 interface regions of interest (ROIs) from biological replicates (>8 haustoria). The average ROI for *T. versicolor* interface transcriptome samples grown on *M. truncatula* was 54,910 μm^2 with a total area of 6.1 million μm^{2}that yielded 144 ng total RNA. This pooled sample had an RNA integrity number (RIN) of 7.6, an A_{260}/A_{280} of 1.58 and an A_{260}/A_{230} of 0.76. The average ROI for *T. versicolor* interface transcriptome samples grown on *Z. mays* was 56,079 μm^2 with a total area of 6.4 million μm^2 that yielded 160 ng total RNA with a RIN of 6.9, a A_{260}/A_{280} of 1.68 and a A_{260}/A_{230} of 0.11.

Linear mRNA Amplification from Laser Microdissected Tissues

The first step of T7 based amplification is cDNA synthesis with an oligo-dT/T7 RNA primer/promoter. It is critical that this step is highly efficient to minimize the bias toward shorter fragment lengths in amplified samples [33]. We frequently observed a yellow-brown material at the host parasite interface that may have contributed to the low initial purity of the interface RNA samples, indicated by the low A_{260}/A_{280} and A_{260}/A_{230} ratios. Thus, we cleaned the interface total RNA with the Zymo™ RNA Clean and Concentrator kit. Subsequently, we observed consistent amplification performance between technical and biological replicates of the cleaned interface total RNA as well as performance consistent with positive control samples of *A. thaliana* young leaf RNA of high quality and purity (28s/18s ratio: 1.9; RIN: 8; A_{260}/A_{280}: 2.0, A_{260}/A_{230}: 2.1).

Amplification of ~100 ng of total RNA routinely yielded 50-100 ug of amplified RNA (aRNA) after two rounds of amplification, which was consistent with the positive control (*Arabidopsis* young leaf total RNA), the expected performance of the Message Amp™ II aRNA kit, and a previous report [34]. The aRNA yield after a single round of amplification was up to 100 ng, which is sufficient for construction of an Illumina sequencing library, yet we chose to amplify the samples for two rounds since it was desirable to have additional aRNA for further analyses including qRT-PCR validation of gene expression profiles. The fragment length profiles, as determined via Bioanalyzer™, were reduced from the first to the second round of amplification, which is consistent with a previous report [33].

Sequencing and Assembly Statistics

Amplified interface RNA samples were sequenced on one lane each of Illumina's Genome Analyzer IIx with an 83 × 83 bp paired-end cycle protocol. Sequencing data are available athttp://ppgp.huck.psu.edu *webcite*[23]. The *T. versicolor* interface transcriptome datasets (Table 1) contained 17.9 million read pairs on *Z. mays* and 19.1 million read pairs on *M. truncatula*. Host reads from each interface transcriptome dataset were mapped to their respective host genomes, leading to the removal of 1.5 million *M. truncatula* reads and 0.4 million *Z. mays* reads from each respective transcriptome data set. Reads were quality trimmed and filtered (see methods), leaving >26 million reads (orphans and mate pairs) for each sample that were then assembled separately using Inchworm (Trinity, [30]) and post-processed to remove exact duplicate or non-translatable sequences. The interface transcriptome assembly of *T. versicolor*grown on *Z. mays* yielded 12.77 Mbp of assembled sequence represented by 28,126 unigenes with an N50 of 525 bp (Table 1). The interface transcriptome assembly of *T. versicolor* grown on *M. truncatula* yielded 12.25 Mbp of assembled sequence represented by 26,709 unigenes with an N50 of 536 bp (Table 1). Sequencing and assembly statistics were similar in all categories (Table 1) indicating that both data sets were of equivalent quality.

Table 1: Read and assembly level statistics for *T. versicolor* interface transcriptomes

Unigene annotation

Host plant		
Sequencing	***Z. mays***	***M. truncatula***
Total raw sequence	2.73 Gbp	2.91 Gbp
Reads Total	**35,894,662**	**38,228,134**
Host reads	(401,352)	(1,588,592)
Host filtered	35,493,310	36,639,542
Quality trimmed	26,947,737	27,325,845
Assembly		
Assembly length	12.77 Mbp	12.25 Mbp
Unigenes Total	**28,126**	**26,709**
Unigenes >500 bp	9,369	9,718
Min/Max length (bp)	197/3,115	197/3,265
N_{50}	525 bp	536 bp
N_{50} >500 bp	731 bp	695 bp
Annotation		
Host unigenes	(4,967)	(7,785)
Non-plant unigenes	(127)	(329)

Triphysaria unigenes	**23,032**	**18,595**
Triphysaria hits	17,887	14,352
Other Plant hits	2,975	2,086
No hits	2,170	2,157

Low quality reads were filtered before assembly, and host sequences were filtered both before and after assembly. Unigenes remaining after removal of host plant and non-plant sequences were aligned with BLASTx to sequences detected in any other PPGP transcriptome library of *Triphysaria versicolor*(http://ppgp.huck.psu.edu/ webcite). Unigenes with less than 95% pairwise identity to either host or to other *Triphysaria* libraries were sorted further if a BLASTx search of the NR (http://www.ncbi.nlm.nih.gov webcite) database yielded alignments of 1e-10 or stronger. The remaining unclassified unigenes were submitted to OrthoMCL DB and InterProScan. Unigenes that remained unclassified after the final screen are called "no hit" unigenes.

Honaas *et al.*

Honaas *et al.* *BMC Plant Biology* 2013 13:9 doi:10.1186/1471-2229-13-9

Unigene Annotation

Unigenes were annotated using an objective classification of known plant genes from the PlantTribes 2.0 database [35,36] as described in Wickett et al. [37]. We assigned genes into a hierarchy of gene clusters, which includes approximate gene families (Tribes), and potentially narrower lineages (Orthogroups) which seek to represent descendants of a single ancestral gene in the collection of reference plant species [35-37]. We also classified unigenes from our experiment using BLAST to query sequence databases (Table1). To identify host derived unigenes in the mixed-species transcriptomes, we established a pairwise nucleotide identity threshold of 95% by querying a collection of *Z. mays* ESTs for *T. versicolor* grown on *M. truncatula* and *vice versa* (Figure S1). The plot of *T. versicolor* unigene identity to each host database is clearly divergent at 95% compared to the reciprocal host database query. To verify that the incident high identity was not due to cross contamination between samples, we also queried the host EST databases with a *de novo* transcriptome assembly of *Lindenbergia philippensis,* a non-parasitic member of the Orobanchaceae [23]. The BLAST identity plot of the *Lindenbergia*transcriptome shows a similar trend to the plot of *Triphysaria* interface transcriptomes queried against the respective non-host databases (Figure S1).

In order to remove host genes that may have escaped detection during pre-assembly read mapping, we further screened the assembly based on sequence similarity to cDNA (Phytozome,[38]) and EST databases (PlantGDB, [39]). This screen removed 4,967 unigenes from the transcriptome of *T. versicolor* grown on *Z. mays* and 7,785 unigenes from the transcriptome of *T. versicolor* grown on *M. truncatula* (Table 1). The same reference transcripts used to screen the raw read data were also used to screen assemblies. The large number of putative host derived unigenes indicates that read screening with Mosaik at default values alone was insufficient to remove all host contamination. After the host screen the remaining unigenes were filtered for *T. versicolor* genes based on sequence similarity to genes detected in other PPGP libraries of*Triphysaria versicolor*[23]. We identified 17,887 unigenes from the transcriptome of *T. versicolor*grown on *Z. mays* and 14,352 unigenes from the transcriptome of *T. versicolor* grown on *M. truncatula* that had >95% identity at the nucleotide level to *T. versicolor* genes from the other PPGP libraries. After removing unigenes with high similarity to *T. versicolor* unigenes from other assemblies in the PPGP database, the remaining 5272 unigenes in the interface transcriptome of *T. versicolor* grown on *Z. mays* and 4572 unigenes in the interface transcriptome of *T. versicolor*grown on *M. truncatula* were used to query the non-redundant protein sequence database (NR) at NCBI [40] using BLASTx at a threshold e-value of 1e-10. Roughly half of the remaining unigenes in each interface transcriptome had best hits to plants including the model species *Arabidopsis, Populus,* and *Vitis,* other Orobanchaceae, or >30 other plant species ("Other Plant Hits" Table 1).

Each *T. versicolor* interface transcriptome had ~2300 unigenes with no significant alignments to sequences in any of the above described external databases (Table 1). We took several additional steps to try to identify these unknown sequences. Though these unigenes are not classified by source, we identified potential plant gene orthologs for ~20-25% of the remaining unigenes via the query of the PlantTribes 2.0 database. We then queried the extensive InterProScan (IPS) [41] and OrthoMCL DB [42] databases with the translated sequences of the remaining, unclassified unigenes. The majority of these unigenes

(>75% in each transcriptome) lacked significant similarity to genes in the OrthoMCL database, nor did they contain IPS peptide motifs (Figure S2); they are thus referred to as "no hit" unigenes (Table 1). The scan of OrthoMCL DB resulted in identification of an additional 100 plant and 7 non-plant unigenes from the *Zea*grown *T. versicolor* and 85 plant and 16 non-plant unigenes from the *Medicago* grown *T. versicolor*(Table 1). Additionally, 370 *Zea* grown *T. versicolor* and 310 *Medicago* grown *T. versicolor* unigenes contained IPS motifs. Roughly half of the OrthoMCL DB hits lack descriptions or have minimal (e.g. one word) descriptions. A similar pattern exists in the IPS search results, where about half of the unigenes with IPS motifs contain only putative secretion signals and/or transmembrane domains. Overall our efforts to classify the unigenes in each assembly resulted in identification of potentially orthologous sequences for 82% of the *Zea* grown*T. versicolor* interface unigenes and 88% of the *Medicago* grown *T. versicolor* unigenes. We were able to assign a putative origin to >90% of unigenes in both transcriptomes and only 5% in each transcriptome remain unclassified. Of these, 493 unigenes from the *T. versicolor* grown on*Medicago* assembly and 536 unigenes from the *T. versicolor* grown on *Zea* assembly are longer than 300 bp and have read support.

The interface transcriptome of *T. versicolor* grown on *Z. mays* and *M. truncatula* contained a total of 127 and 329 unigenes, respectively, with best hits to non-plant species (Table 1). The non-plant component of each interface transcriptome included best hits to 16 taxa shared by both interface libraries. These included *Escherichia, Aspergillus, Clavispora, Burkholderia*, and others. Among this set, *Burkholderia* was the most highly represented taxon (>20 fold increase over any other species) in the non-plant component of both interface transcriptomes. These interface *Burkholderia*sequences were not detected in the reference transcriptome (TrVeBC1, sequence identity cutoff >90%).

The remaining "no hit" sequences, especially those >300 bp with read support, could represent unannotated host or parasite genes, uncharacterized associated symbionts, or incidental contamination. All remaining unigenes not assigned to a host plant or non-plant source in the NR database (23,032 for *T. versicolor* on *Z. mays* and 18,595 for *T. versicolor* on *M. truncatula*) are considered collectively as "putative *T. versicolor* derived unigenes."

Comparative Interface Transcriptome Profiles of *T. Versicolor*

To examine the profiles of the *T. versicolor* interface transcriptomes, we used an annotated transcriptome from above ground tissues of autotrophically grown *T. versicolor* as a reference ([23]*Triphysaria* assembly TrVeBC1). We sorted unigenes by (PlantTribes 2.0) Orthogroups to identify host-specific and shared components of the interface transcriptomes of *T. versicolor* grown on *Z. mays* and *M. truncatula* (Figure2). As expected, the largest number of Orthogroups (5947, or 53.6% of the total detected in *T. versicolor*) were shared between all three transcriptomes, and likely represent expression of genes involved in processes common to a wide variety of cell types. A large number of Orthogroups (1124) were shared between the interface transcriptomes of *T. versicolor* interacting with both hosts. These genes likely include a putative core set of parasite genes that are active irrespective of the host plant species. Many additional Orthogroups were either exclusive to the interface and host-specific (677 for *Z. mays* and 361 for *M. truncatula*), or shared with above ground phases of growth (1066 for *Z. mays* and 314 for *M. truncatula*).

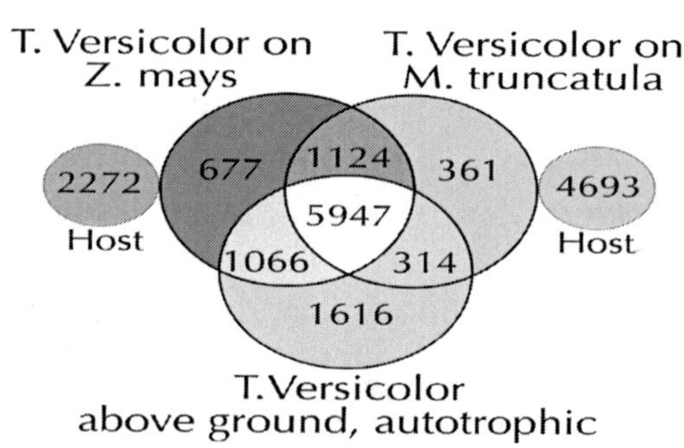

Figure 2: Transcriptome Orthogroup Venn. Venn diagram showing the number of Orthogroups in the interface transcriptomes of *T. versicolor*with hosts *Z. mays* and *M. truncatula* and an above ground, autotrophically grown *T. versicolor* transcriptome (TrVeBC1) constructed from leaves, stems and inflorescences. Also shown are the numbers of host-derived Orthogroups. The lack of overlap between host and parasite transcriptomes does not imply lack of shared Orthogroups, but indicates the total number of host Orthogroups

for a point of comparison.

Our annotation strategy includes assignment of a GO Slim category term derived from the best BLAST hit in PlantTribes 2.0. GO Slim categories are the broadest designations of GO and are useful for transcriptome-wide comparisons. The host-specific component of the *T. versicolor*interface transcriptome is likely to contain genes that interact with unique aspects of host biology while those that are shared likely contain genes essential for parasitism. To determine if the annotation profiles were similar between the overlapping and unique transcriptome components we plotted the proportion of GO Slim categories of unigenes (Figure3) represented by unique or overlapping Orthogroups in Figure2. GO Slim category profiles between equivalent components of each interface transcriptome were generally similar to each other, yet often distinct from profiles of non-equivalent components of each interface transcriptome (Figure3). For instance, interface-unique Orthogroup profiles were similar between both interface transcriptomes, yet distinct from the above ground Orthogroup profiles of both interface transcriptomes.

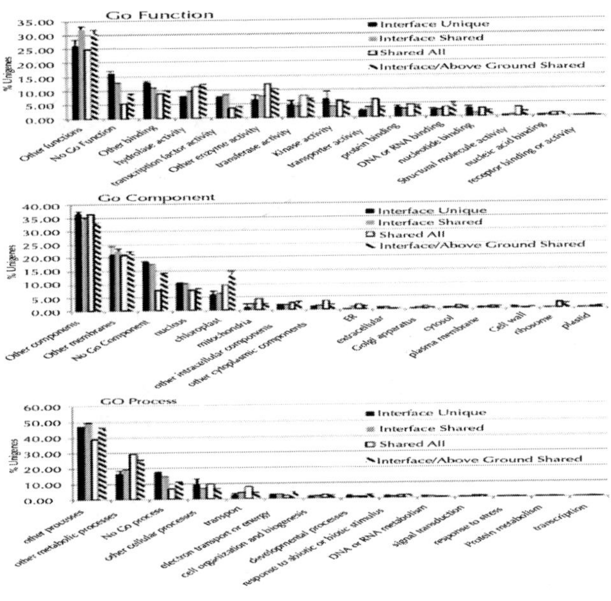

Figure 3: GO Slim category summary. GO Slim category terms of unigenes in interface transcriptomes of *T. versicolor* and the above ground reference

assembly of *T. versicolor.* Each series displays the average number of unigenes in equivalent transcriptome components with a given GO Slim term. For instance, "Interface Unique" indicates the average number of unigenes from interface unique components in both*Medicago* and *Zea* grown *T. versicolor* transcriptomes. Error bars are standard error of the mean. "Interface Unique" = unigenes from Orthogroups that are host and interface specific, "Interface Shared" = unigenes from Orthogroups that are interface specific and shared between interface transcriptomes, "Shared All" = unigenes from Orthogroups shared between both interface transcriptomes and the above ground transcriptome, "Interface/Above Ground Shared" = unigenes from Orthogroups that are shared between the above ground, autotrophic transcriptome and the host-specific interface transcriptome.

GO Slim category profiles in overlapping and unique sets of Orthogroups within each interface transcriptome were tested for proportionality by a Chi-Square test (Figure S3A-F). The results of all 6 tests showed disproportionality and were strongly significant (P<<0.0001). The number of unigenes in each GO Slim category, with strong residual values (strongly positive or strongly negative, thus disproportionate), is indicated in Figures S3A-F. The results of this analysis are concordant with the plot of GO Slim category profiles (Figure 3). The most striking result is the consistently strong over-representation of unigenes lacking GO Slim categories in interface-unique Orthogroups compared to the consistent weak representation of unigenes lacking GO Slim categories in the shared Orthogroups (S3A-C: column D, Figure S3D-F: column H). Interestingly the Orthogroups shared in the interaction of *T. versicolor* with both hosts are overrepresented by "transcription factor activity" GO Slim Function terms (Figure S3A: column B and 3D: column F) and underrepresented by "transport" GO Slim Process terms (Figure 3: column B and 3F: column F). This indicates that there are transcription factor genes active at the parasite-host interface that are not expressed in the above ground reference transcriptome. In contrast, the transporter gene families expressed at the interface are expressed in the above ground reference transcriptome as well.

Highly Expressed Genes at the Host-Parasite Interface

An advantage of using a *de novo* assembled transcriptome for RNA-Seq is an intrinsic threshold for transcript detection. If the transcript is represented by sufficient reads for *de novo* assembly, the presence of a target for *de novo* RNA-Seq is evidence for the presence of a transcript. The reference assembly TrVeBC2 [23] includes data from the haustorium of *T. versicolor* grown on *M. truncatula* and was used as a reference to map reads from each interface transcriptome. We correlated normalized reads (reads/kilobase/million mappable reads (RPKM)) from unigenes belonging to Orthogroups shared between the interface transcriptomes and the above ground reference transcriptome, TrVeBC1 (Figure S4). For unigenes detected in both interface transcriptomes the correlation was high (Pearson's R= 0.81), which indicates low technical and biological variability between the interface transcriptomes.

To determine the expression level of each unigene we also mapped reads to each respective interface *de novo* assembly. The 20 most highly expressed unigenes in each set of shared and unique Orthogroups from the two transcriptomes are presented in Figures S5A-C. We queried this set of 120 unigenes against the NR database using BLASTx (e-value threshold: 1e-10) and 17 annotated plant genomes using BLASTn and BLASTx. The results of the database queries using BLAST are presented in Figures S5A-C. The best-hit descriptions from searches in NR were concordant with the annotations assigned using PlantTribes 2.0. We also used InterProScan to predict signal peptides and transmembrane domains for the unigenes listed in Figure 5A-C. The motif prediction tools frequently identified putative transmembrane domains in unigenes annotated as transporters and secretion signals in unigenes annotated as secretory proteins. Of the 120 unigenes listed (Figures S5A-C), nine had no hit when queried against NR. The remaining unigenes had best hits to plant species. About 30% of these 120 unigenes have either no BLAST hits in NR or align to predicted, hypothetical, or otherwise uncharacterized sequences. This result is consistent our finding that the interface is enriched with unigenes that lack GO Slim category assignments (thus functional annotations).

Among the most highly expressed genes in the shared orthogroups of interface samples of *T. versicolor* grown on in both *Z. mays* and *M. truncatula* (Figure S5A) are a β-expansin gene (see below), genes for several other cell wall modifying enzymes, and a gene encoding a putative ap2-erf domain transcription factor. A striking pattern in the shared interface Orthogroups was 10 unigenes (including six of the most strongly expressed unigenes from *T. versicolor* grown on *Medicago*) with sequence identity to annotated pathogenesis-related proteins in other eudicot species. A single *M. truncatula* unigene passed through the host plant removal process in the common interface component (ID 5537); this also shared high sequence identity with a pathogenesis-related protein.

Of the unigenes listed in Figures S5A-C, 42 from the interface transcriptome of *T. versicolor* grown on *Z. mays* and 34 from the interface transcriptome of *T. versicolor* grown on *M. truncatula* had strongest BLAST hits to Asterid genomes (including *Mimulus guttatus*). When we queried the 17 plant genomes database there were slightly more best hits to legumes in the*Medicago* grown *Triphysaria* data set, perhaps because there is less sequence divergence between the eudicots *Triphysaria* and *Medicago* than between the more distantly related *Triphysaria* and*Zea*. This results in a somewhat broader range of ambiguous sequence identity between host and parasite. Despite rigorous filtering, a single putative *Z. mays* transcript and three putative *M. truncatula* transcripts persevered (indicated in bold) in the highly expressed gene list in Figure S5.

A Novel β-Expansin is Differentially Expressed at the Parasite-Host Interface

Among the highly expressed unigenes observed in the interface transcriptome of *T. versicolor*grown on *Z. mays* was a putative β-expansin (Figure S5A). Manual curation of the read mapping data indicated that it was highly expressed when grown on *Z. mays*, and a nearly identical unigene from the *M. truncatula*-grown *Triphysaria* interface was lowly expressed. This apparently host-specific gene expression pattern was of interest because expansins are cell-wall loosening proteins (for a review, see [43]) that have been implicated in the interaction between parasitic plants and their hosts [31,32]. While the β-expansin gene was expressed in both samples, read mapping evidence suggested that this gene (unigene 772) was highly differentially expressed. As a point of comparison, we investigated a putative α-expansin, unigene 11, which showed a reciprocal pattern of high expression in the interface transcriptome of *T. versicolor* grown on *M. truncatula*. We verified the nucleotide sequence of unigenes 772 and 11 via dye-terminator sequencing of PCR products amplified from interface aRNA.

Phylogenetic analysis of β-expansin unigene 772 shows that it is nested within a supported clade of dicot β-expansin sequences (Figure S6A) indicating that unigene 772 is a dicot β-expansin and not a *Z. mays* derived sequence. Annotation via InterProScan supports an expansin identity for 772 and shows a putative 5′ signal peptide (Figure S5A), consistent with a role in the apoplast that is typical for expansins. Additionally, the results of all of the BLAST searches suggest that unigene 772 is a *T. versicolor* derived sequence. Phylogenetic evidence for unigene 11 does not yield a well-resolved tree of α-expansins (Figure S6B), but the BLAST results suggest that the pairwise nucleotide identity to known, or putative (e.g. ESTs) *M. truncatula* genes is <70%, while unigene 11 has high identity (>95% pairwise nucleotide identity) to *Triphysaria* unigenes in other PPGP assemblies.

Quantitative Real-Time PCR Verification of Host Specific Expansin Expression

We sought to verify the reciprocal expression patterns of these two expansins via qRT-PCR. Unigenes 772 and 11 were assigned formal names TvEXPB1 and TvEXPA4, respectively. We verified that primers were specific to their targets by melting curve analysis. To further verify that the TvEXPB1, TvEXPA4, and TvActin primers were specific to parasite transcripts, we harvested portions of host roots that were immediately adjacent to mature *T. versicolor* attachments and interrogated them via qRT-PCR. In these host root samples we were able to detect ZmActin in *Z. mays* root samples and MtActin in *M. truncatula* samples, while the parasite primers yielded signal consistent with background.

We interrogated biological replicates of the *T. versicolor* host-parasite interface cells grown on bothZ. *mays* and *M. truncatula* via qRT-PCR for expression of the reference gene, TvActin, TvEXPA4, and TvEXPB1 (Figure4). TvEXPB1 is up-regulated >120 fold (P=0.024) in *T. versicolor* haustorial interface cells grown on *Z. mays* relative to *T. versicolor* grown on *M. truncatula*. TvEXPA4 shows a weak reciprocal pattern (P=0.17). The expression patterns of TvEXPA4 and TvEXPB1 are concordant with our *de novo* RNA-Seq results. Additionally, when we examined TvEXPB1 expression in whole haustorium samples the signal was indistinguishable from background, suggesting that the massive upregulation of TvEXBP1 is specific to a small number of interface cells.

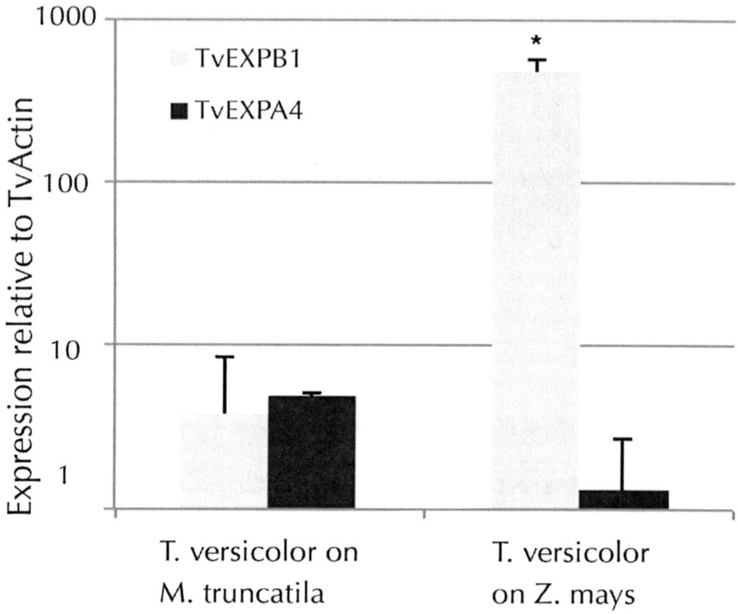

Figure 4: Differential expansin expression. qRT-PCR analysis of TvEXPA4 and TvEXPB1 expression relative to TvActin in parasite-host interface cells harvested by LPCM from the haustoria of *T. versicolor**P<0.05.

DISCUSSION

Using a workflow that allowed us to sample, sequence, and *de novo* assemble transcriptomes from cells at the host-parasite interface we have shown that *T. versicolor* expresses genes in a host specific manner. This preliminary look at genes expressed at the parasitic plant-host plant interface suggests that the basis for generalist parasitism is constituted, at least in part, by host-specific patterns of gene expression. Generally, this work demonstrates the potential to discover genes *de novo* and examine genome-wide patterns of gene expression in a highly tissue-specific manner in organisms that lack a sequenced genome.

Laser Microdissection (LM) is a Powerful Tool for Plant Transcriptomics

The power to develop a comprehensive picture of any biological system lies with understanding the myriad processes underway in complex organs and tissues. A primary hurdle to revealing this picture is the ability to identify, separate, capture and analyze tissues and cells of interest. Several authors have emphasized the importance of high resolution, high through-put investigations of gene expression in a tissue- and cell-specific manner as well as the need to survey gene expression in a global manner, and why LM (including LPCM) is emerging as a powerful tool for genomics [44-47].

LM generated samples from some model systems have been examined using microarrays [48-53] allowing investigators to analyze global gene expression patterns in specific tissues and cell types. More recently, LM has been increasingly coupled with NGS to sequence the transcriptomes of various tissues in *Z. mays*[54-56] and *S. lycopersicum*[34]. The advent of *de novo* transcriptome assembly now makes global surveys of gene expression in specific tissues and cells of non-sequenced organisms a logical next step. To examine the parasite-host interface transcriptome of*T. versicolor,* we combined LPCM with robust T7 based linear RNA amplification [33,57-60] in concert with Illumina mRNA-Seq, high performance *de novo* transcriptome assembly [30], and various assembly post-processing tools.

For the haustorium of *T. versicolor* our sampling strategy was based largely on detailed histology and transmission electron microscopy work done with field-collected specimens [12,13]. This critical background information allowed us to identify cells of interest within the haustorium and then subsequently identify regions of the haustorium in cryosections that contained these cells. Plant tissues must be embedded prior to LPCM and preparatory steps can have an impact on the quantity and quality of RNA preparation [61]. However, the ability to identify tissues of interest must be balanced with downstream usability. The histological quality of the section is an important consideration that may determine the sample preparation method. Paraffin embedded sections generally provide high histological quality at the expense of RNA quality and yield [62,63]. Histological quality increases with thinner sections for sampling at a finer spatial resolution and the efficiency of pressure catapulting increases with thinner sections, yet Kerk et al. [64] report increased RNA yield from thicker sections. We found that the optimum section thickness for capturing interface cells of the *T. versicolor* haustorium was 20-25 μm. This was determined based on a balance of our ability for histological identification of tissues and cells of interest with efficient tissue release from the slide during the pressure catapult phase of LPCM. The integrity of plant tissues that are susceptible to damage by flash freezing can be preserved by infusion with a cryoprotectant [65]. Cryosectioning with the CryoJane™ (a cryosection transfer system) allowed us to easily capture serial cryosections of *T. versicolor* haustorium that routinely yielded high quality RNA from carefully chosen samples.

By design, our sampling strategy minimized the likelihood that differences in gene expression arose from temporal or spatial sampling artifacts. The co-culture of *T. versicolor* is not highly synchronous, so our sample of haustoria represents a broad temporal window of connections that are ~8-10 days old. We collected ~110 interface ROIs which diminishes the likelihood of spatial sampling artifacts. Furthermore, highly similar statistics throughout sample processing and data analysis, including a high correlation of read counts for unigenes in shared Orthogroups (Figure S4) and verification of the expression pattern of TvEXPB1 in additional experiments with biological replicates (Figure 4), indicates low variation from either biological or technical sources.

Parasite-Host Interface Transcriptomes Help to Identify Core Parasite Genes

Orobanchaceae include the pernicious weeds *Striga*, *Orobanche* and *Phelipanche*, as well as the model parasite *Triphysaria*. We reasoned that by discovering a subset of genes that likely include those central to the response of *T. versicolor* to distantly related hosts, we would move closer to identifying a core set of parasitism genes operating in the weedy Orobanchaceae. The molecular dialogue of the parasitic plant with its host is largely uncharacterized [66]. Here we report the identification of the subset of genes expressed at the host-parasite interface by *T. versicolor* in response to both of the distantly related hosts *Z. mays* and *M. truncatula*. Additionally, we identified host-specific patterns of gene expression indicating that the generalist parasitic plant *T. versicolor* maintains suites of genes for use with different hosts.

The genes of *T. versicolor* expressed at the host-parasite interface in response to both *Z. mays*and *M. truncatula* included a substantial set of genes annotated in other plant species as pathogenesis- (or pathogen-response) related proteins. This included six of the most highly expressed genes in the *Triphysaria* interface grown on *M. truncatula*, and three when grown on *Z. mays*, but a number of additional unigenes were also putative homologs of genes that are upregulated during pathogen invasion (dirigent-like, acidic endochitinase, disease resistance proteins, etc.). While upregulation of genes of these classes would be expected as a defense response to pathogens, this observation suggests that pathways commonly involved in plant protection are also turned on by the parasite during the process of host invasion. It has been already suggested that parasitic plants may have recruited pathways from allelochemical detoxification for use during haustorium signal transduction [67]. Whether *Triphysaria* is defending itself during the invasion process with pathogen resistance (PR) genes, or has recruited PR pathways as offensive 'weapons' is not yet clear, but it does appear that a portion of the core gene set for *Triphysaria* parasitizing either host has been derived from pathways that plants use for defense against pathogens.

There is no consensus for why true generalist parasites occur or persist, even infrequently, despite evidence for an evolutionary trend toward specialization ([18] and references there-in). We hypothesized that a successful generalist feeding strategy might rely heavily on general-purpose genes that were deployed for feeding on any host. Instead, we see large components of the interface transcriptome that are detected only when feeding on one of the tested host plants. Of the 2162 Orthogroups detected only in interface transcriptomes, about half (52%) were expressed on both host species, but a substantial fraction of interface Orthogroups were detected only when grown on maize (31%), or on *Medicago* (17%). Long-term maintenance of extensive genetic machinery directed toward parasitism of subsets of possible hosts will require maintenance of selective pressures on the genes that are deployed in a host-specific manner. Unless the parasite routinely encounters and parasitizes a wide range of host plants, selective pressure on some host-specific genes will be relaxed and the gene functions eventually lost, limiting the plant's potential to parasitize some hosts, despite the potential advantages of a large host range. The evidence for host specific gene expression suggests that *T. versicolor* regularly parasitizes across a broad host range, maintaining selective pressure on genes used in a host-specific manner. Indeed the advantage of a broad host range has been explored where host populations are moderately variable in space and time [19]. This study serves as a starting point for a broader survey of putative and potential hosts for *Triphysaria* as well as a first look at gene expression profiles of highly specialized tissues in parasitic plant haustoria.

It is intriguing that the host-specific and shared-interface gene families (Orthogroups) were over-represented by unigenes with no GO Slim category term assignment compared to the above ground transcriptome. Because Orthogroups were defined based on a classification of sequenced and annotated plant genomes [35-37] to which the parasite genes were assigned, this observation highlights the fact that genes expressed in the haustorium include many that have been recruited from the subset of genes whose function is not yet known in any plant. In addition, approximately 1300 unigenes from each interface transcriptome lack strong homology to any known sequence, though 25% of these unigenes are high identity reciprocal best hits in the interface data sets from each host. Further, because such patterns are reproducible in the interface transcriptomes of *T. versicolor* when grown on different host plants, these data suggest that genes of unknown function are expressed in the haustorium in a host-specific manner. Our data also suggest that underground phases of growth in *T. versicolor* are enriched for genes of unknown function.

Of those unigenes with GO Slim category assignments, the shared interface-specific Orthogroups are overrepresented for the GO Slim Function category "transcription factor" and underrepresented for the Go Slim Process category "transport." This indicates that there are transcription factor Orthogroups unique to the interface (relative to the above ground reference transcriptome), yet Orthogroups involved in transport processes are active in all three transcriptomes examined. The latter observation regarding transport does not rule out differential expression of particular genes that are expressed in all three transcriptomes in this study.

Tvexpb1 Encodes A *T. Versicolor* Β-Expansin that is Part of a Host-Specific Response

The expansin gene family includes 4 main groups: α, β, expansin-like A and expansin-like B [43]. Expansins are thought to loosen plant cell walls by allowing slippage of cell wall polymers [43]. Their activity is non-enzymatic, but they are distantly related to glycoside hydrolase family proteins [43]. Expansins are involved with cell growth [68] and have been implicated in the interaction of bacterial plant pathogens [69,70], plant-parasitic nematodes [71,72] and parasitic plants [31,32] with their plant hosts. In each each case, expansins are suggested to play a role in host invasion.

Expansin activity has been assayed in the cell walls of both monocots [73-75] and dicots [76, 77]. β-expansins have activity that is specific to the cell walls of grasses, but not dicot or most other monocot cell walls [78]. The reverse pattern of action is found for α-expansins, which suggests substrate specificity for both α- and β-expansin proteins [43,79]. Typically, expansins are found at low concentrations [43] with the exception of grass pollen that secretes massive amounts of β-expansins [76] that likely serve to loosen stylar tissue during pollen tube growth [80,81]. Although the exact mechanism of action remains unknown, grass cell walls have relatively small amounts of xyloglucan and pectin; these are replaced with β-(1→3), (1→4)-D-glucan and glucuronoarabinoxylan. Both of these grass cell wall components are potential targets of β-expansins in their wall-loosening activity [82]. Throughout more than a decade of research, the accumulation of β-expansins to high levels was thought to be specific only to grass pollen [82].

In this study, we have shown that the transcript level for the β-expansin TvEXPB1 is among the most highly expressed genes in the interface transcriptome of *T. versicolor* grown on *Z. mays*. Relative to the interface transcriptome of *T. versicolor* grown on *M. truncatula,* the expression of TvEXPB1 is greatly up-regulated (>120-fold) in the parasite-host interface tissues of *T. versicolor* grown on *Z. mays*. The massive and host-specific expression of TvEXPB1 is suggestive of a role in a dicot parasite's interaction with a grass host. Additionally, the signal for TvEXPB1 in whole haustorium samples of *T. versicolor* grown on *Z. mays* was undetectable relative to the interface samples suggesting that TvEXPB1 is highly specific to the host-parasite interface.

Taken together, the evidence for grass cell wall-specific activity of β-expansins and the massive upregulation of a parasite β-expansin at the parasite-host interface suggests that *T. versicolor*expresses TvEXPB1 when interacting with the monocot host *Z. mays* in an effort to manipulate host cell walls. Heide-Jørgensen and Kuijt [12,13] observed that host cortical and epidermal cells seemed crushed and displaced at the host parasite interface in the haustorium of *T. versicolor.* The mechanism of host tissue displacement may include cell wall modifying proteins like expansins that have host cell wall-specific activity. Such a mechanism would allow the parasite to soften or separate host cell walls without affecting the integrity of its own cell walls in the penetrating haustorium. The specific role that expansins play in the host parasite interaction will only be uncovered through detailed functional analysis. This includes focused gene expression analysis, targeted silencing of *T. versicolor* expansin genes and biochemical characterization to determine the substrate specificity of the expansin proteins encoded by these genes.

Are a Ubiquitous Genus of Soil Bacteria Symbionts of *T. Versicolor?*

The best represented genus in the non-plant transcriptome component (~60% of non-plant hits in*T. versicolor* grown on *Z. mays* and ~45% non-plant hits in *T. versicolor* grown on *M. truncatula)*was *Burkholderia,* a common genus of soil bacteria. The high frequency of hits to this bacterium is surprising for three reasons: (1) the relative frequency of other non-plant genera was much lower, (2) *T. versicolor* seeds were aggressively surface sterilized prior to axenic co-culture, and (3)*Burkholderia*-derived unigenes were not detected in the above ground reference assembly. The low frequency of hits to other genera (including plant pathogenic fungi, human and other bacteria) could be explained by incidental contamination from the lab environment, however the preponderance of *Burkholderia* hits only in the interface samples suggest the presence of an organism belonging to this genus in the co-culture system. The unigenes were generally <1 kbp (indicating transcript-sized unigenes) and RNA samples were DNase treated, diminishing the likelihood that these unigenes originated from genomic DNA contamination.

The genus *Burkholderia* has received increasing attention in the last two decades due in part to a diverse catalog of host interactions ranging from human pathogen to plant-growth promoting rhizosphere fauna [83]. Of particular interest here is their potential role as beneficial plant endosymbionts [83]. Attsat [84] noted the presence of filamentous bacteria-like structures in haustorial cells and that application of terramycin significantly reduced haustorium formation in*Orthocarpus purpurascens* (syn. *Castilleja exserta*), a hemi-parasite that is a close relative of *T. versicolor*[85]. The presence of *Burkholderia* in an axenic co-culture system points to an intriguing possibility: a species of *Burkholderia* was carried through surface sterilization with the seeds of *T. versicolor.* Evidence that may suggest a role for a prokaryotic endosymbiont in parasitic plant biology [84] considered with evidence for the presence of a genus of beneficial soil bacteria hints at a symbiosis between *Burkholderia* and *Triphysaria*. The presence and possible roles of*Burkholderia* at the host-parasite interface in *T. versicolor* are currently under investigation.

CONCLUSIONS

Triphysaria represents an important asset in the identification of genes and evolutionary processes that are central to parasitic plant biology, and one that is key to the development of new control strategies for the weedy Orobanchaceae. Highly tissue-specific transcriptome analysis in*Triphysaria versicolor,* an experimental model parasitic plant that has yet to have its genome sequenced, has revealed host-specific gene expression. Furthermore, genes represented at the parasite-host interface are enriched for genes of unknown function relative to above ground phases of growth. Ongoing development of molecular techniques, including parasitic-plant transformation [8,86] will facilitate the functional characterization of genes whose roles may be central in the parasitic plant-host plant interaction.

METHODS

Growth of Plant Material

Triphysaria Versicolor

Seed was collected from hundreds of open pollinated plants growing in coastal grassland stands of Napa California laboratories (University of California, Davis, USA) from the same source population as our prior transcriptome sequencing studies [37,87]. Seeds were surface sterilized in 70% ethanol for 10 min. while gently shaking, and then washed 3x with sterile distilled H_2O. Seeds were further sterilized and scarified by a wash with a 50% bleach + 0.01% Triton X-100 (Sigma) solution for 30 min. while gently shaking. Seeds were then washed 10x with sterile distilled H_2O and placed in petri dishes containing co-culture medium (1/4x Hoagland's basal salt and nutrient mix, 7.5 g/L sucrose, 6 g/L plant tissue culture grade agarose, pH of 6.1) and wrapped with Parafilm™. Seeds were stratified for 4 days at 4°C in the dark then transferred to a 16°C growth chamber under a 12-hour light regime with a light intensity of 30 µmoles photons/m²/sec. *T. versicolor* seedlings were grown for 14–17 days then transferred to fresh co-culture plates with hosts.

Medicago Truncatula (A17)

Seed was generously provided by Zengyu Wang (Noble Foundation, Oklahoma, USA). Seeds were scarified by incubation with occasional stirring in 18M H_2SO_4 for 8 min., and then washed 5× with sterile distilled H_2O. Seeds were surface sterilized by a wash with 50% bleach + 0.01% Triton X-100 (Sigma) for 3 min. while gently shaking. Seeds were washed 10x with sterile distilled H_2O and placed in a 50 mL conical bottom tube with 25 mL of sterile distilled H_2O at 25°C, in the dark, over-night while gently shaking. The next day seedlings were transferred to sterile filter paper (Whatman #5), moistened with sterile distilled H_2O, and placed in petri dishes. Seedlings were placed at 25°C, in the dark, over-night. Seedlings were transferred to co-culture medium and wrapped with Parafilm™ then placed in a growth chamber at 25°C under a 16-hour light regime with a light intensity of 100 µmoles photons/m²/sec for 5–7 days, then transferred to fresh co-culture plates with parasites.

Zea Mays (B73)

Seed was generously provided by David Braun (University of Missouri, Columbia, USA). Seeds were surface sterilized by a wash with 20% bleach+0.08% Triton X-100 (Sigma) for 10 min. while gently shaking. Seeds were washed 3x with sterile distilled H_2O. Seeds were further surface sterilized by a wash with 70% ethanol for 5 min. while gently shaking. The ethanol was decanted and the seeds were left to air dry in a laminar flow hood. Dry seeds were placed in petri dishes containing co-culture medium and wrapped with Parafilm™ then placed in a growth chamber at 25°C under a 16-hour light regime with a light intensity of 100 µmoles photons/m²/sec for 5–7 days, then transferred to fresh co-culture plates with parasites.

Parasite and Host Co-Culture

Zea mays or *Medicago truncatula* were transferred to fresh co-culture medium at the times indicated above. Host roots were carefully oriented to allow placement of *T. versicolor* (grown as described above) seedlings in close proximity to host roots, with the parasite root tips 0.5-1 mm from the host root.Co-culture plates were sealed with Micropore™(3M) surgical tape and placed in a growth chamber at 25°C under a 16-hour light regime with a light intensity of 100 µmoles photons/m²/sec for 8–10 days.

Tissue Processing and Sample Preparation for Sequencing

Tissue Harvest

After 8–10 days co-culture, the haustoria of *T. versicolor* were harvested by making cuts in both the host root and parasite root ~1-2 mm adjacent to the haustorium under a stereomicroscope and were embedded in Shandon Cryomatrix™ (Thermo Scientific) dispensed in a 10 mm x 10 mm x 5 mm Cryomold™ (Tissue-Tek). Dissected tissue was quickly oriented with the host root on the vertical axis and with all haustoria equidistant from the bottom of the mold. Samples were quickly frozen on dry ice and stored at -80°C.

Cry sectioning and Dehydration

Embedded haustoria were sectioned in a Cryotome SME (ThermoFisher Scientific) at 20-25 µm. Frozen sections were mounted to slides using the Cryo-Jane™ Tape Transfer System (Leica Microsystems). Mounted sections were dehydrated in a series of organic solvent baths (RT 70% ethanol (10 min.), 4°C 70% ethanol (2 min.), 4°C 95% ethanol (2 min.), 4°C 100% ethanol (2 min.), 4°C 100% xylene (2 min.), 4°C 100% xylene (2 min.), and finally RT 100% xylene (2 min.)). Slides were allowed to dry in a fume hood for 30 min.

Laser Pressure Catapult Micro-Dissection (LPCM)

The P.A.L.M. Microbeam™ System (Zeiss) was used to harvest cells of interest. Only the pressure catapult function was used and the laser energy setting for the pressure catapult function was minimized. The laser focus was fixed, but the objective focus was manually adjusted to ensure efficient removal of tissues of interest during LPCM. Dissected tissue was captured in opaque adhesive cap 0.5 mL tubes (Zeiss, part #415101-4400-250) and stored at -80°C.

RNA Extraction and RNA Cleanup

Total RNA was extracted from LPCM harvested material using the PicoPure™ RNA isolation kit (Arcturus) with adaptations derived, in part, from documentation accompanying the opaque adhesive cap 0.5 mL tubes (Zeiss, part #415101-4400-250). The changes include modification of the RNA extraction step as detailed in the Arcturus protocol (Step 1) and are as follows: We removed the adhesive cap and placed it into the tube body of a 0.5 mL Eppendorf Safe-Lock™ tube (part # 022363719) to achieve a better tube-cap seal. 50 μL of buffer XB was added to the tube and it was vortexed inverted for 30 seconds. The tube was then placed in a temperature equilibrated (42°C) custom clamping device (not shown) to prevent leakage during the lysis step, in which the tube is inverted and incubated at 42°C for 30 min. in an air incubator. At 10 min. intervals the entire clamping device containing the tubes was vortexed inverted for 30 seconds. After the adapted lysis step we followed the PicoPure™ RNA isolation kit protocol beginning at step 2, RNA isolation. RNA was DNase treated on-column according to Appendix A in the PicoPure™ RNA isolation kit protocol. Total RNA was assessed on the Agilent Bioanalyzer using the RNA 6000 Nano kit (Agilent) with the Plant Total RNA assay with specific attention to the RNA Integrity Number (RIN, scale of 1 (degraded) to 10 (intact)) [88]. An additional clean-up of the total RNA prep was required to remove what we suspected to be poly-phenolics and secondary metabolites that interfered with downstream enzymatic treatments. The RNA Clean and Concentrator™-5 kit (Zymo Research) was used to clean and concentrate the total RNA extracted from LPCM harvested samples following the General Procedure in protocol version 2.0.6 and RNA was eluted in 10 μL RNase-free water.

T7 Based RNA Amplification of Mrna

The Message Amp™ II aRNA kit (Ambion) was used to amplify the poly-A RNA contained in total RNA samples to yield amplified RNA (aRNA). The input amount was approximately 100 ng of total RNA. The manufacturer's instructions were followed and the first round of *in vitro* transcription (IVT) was allowed to progress for 14 hrs. The entire aRNA sample was concentrated in a Speed-vac (Savant) to <10 uls then entered into second round cDNA synthesis. The second round IVT was allowed to progress for 4 hours. Total RNA was assessed on the Agilent Bioanalyzer using the RNA 6000 Nano kit (Agilent) with the mRNA assay. Typical yields after the first round of amplification were up to 100 ng aRNA and yields after the second round of amplification ranged from 50-100 ug aRNA. High quality Arabidopsis young leaf RNA was used as a positive control and RT-PCR grade water (Ambion, included with the Message AmpII kit) was used as a negative control for amplification.

Illumina Paired-end Library Construction

aRNA was used as input in Illumina's mRNA Seq library prep protocol (Rev D). We omitted the poly-A selection step and moved directly to "Fragment the RNA." 100 ng aRNA was fragmented and the manufacturer's instructions were followed for the rest of the library preparation with the following exceptions: 1) we size selected the adapter-ligated library fragments at 300 bp rather than 200 bp at the "Purify the cDNA Templates" step, 2) we performed a second size selection/purification step by running a gel in a similar fashion as described in the "Purify the cDNA Templates" step and excised the band at approximately 325 bp that contained the products of library enrichment. The second size selection was done to purify the library and further constrain the fragment distribution as recommended by Illumina for paired-end mRNA Seq. The Illumina RNA Seq library was assessed on the Agilent Bioanalyzer using the DNA 1000 kit (Agilent).

Sequencing, Bioinformatics and Phylogenetics

Sequencing

Paired-end (83×83 bp) sequencing was performed on the Illumina Genome Analyzer 2x by the Genomics Core facility at the University of Virginia in Charlottesville, VA USA. Each library was sequenced in one lane.

Post Sequencing Data Processing and Annotation

Contaminating sequences were removed from the pre-assembled, paired-end reads by alignment to the annotated coding DNA sequences of *Medicago truncatula*[38] and *Zea mays*[38] genomes using version 1.1.0014 of Mosaik Assembler [89] with the recommended parameters (hs = 15, mm = 12, and act = 35). Unaligned reads were then trimmed to remove low-quality bases (<Q20) from the ends using the quality trim program of CLC Assembly Cell version 3.2 (http://www.clcbio.com/index.php *webcite*) requiring additionally that the remaining

read fragment be at least half the original read length. Paired-end read files were reconstructed from the trimmed read fragments and orphaned single-end read fragments written to separate files using a custom script. The resultant filtered and trimmed set of reads for each library was then *de novo* assembled using the Inchworm component in release 03122011 of Trinity [30] with default parameters. Assemblies were filtered using version 2 of ESTScan [90] to remove sequences that had numerous frame-shift errors in coding regions, and version 4.0 of USEARCH [91] to remove similar (sub)sequences (to create non-redundant *de novo* assemblies). The post-processed unigenes for each build were then queried (BLASTx, 1e-5) against the 10 genomic proteomes in PlantTribes 2.0 [36,37] and assigned to PlantTribes 2.0 Orthogroups based on the cluster containing the best BLAST hit. Expression levels for each unigene was determined by mapping reads back to *de novo* assemblies (each interface transcriptome as well as TrVeBC1 and TrVeBC2) and computing RPKM value with the *High-throughput Sequencing RNA-Seq Analysis* program of CLC Genomics Workbench version 4.6 (parameters: mismatch cost = 2, insertion cost = 3, deletion cost = 3, length fraction = 0.5, similarity = 0.8, min insert size = 100, and max insert size = 250).

Post Annotation Assembly Filtering

To determine the appropriate pairwise identity (at the nucleotide level) to reference host sequences, BLASTn was used to determine the frequency of unigene pairwise identity to reference host ESTs (*Zea mays* and *Medicago truncatula*, mRNA ESTs [39]). The interface transcriptomes of*T. versicolor* grown on *Z. mays* and *M. truncatula* were BLASTed into both the *Z. mays* and *M. truncatula* EST databases. The BLAST to the non-host databases was to determine the incident nucleotide pairwise identity of unigenes to a set of host references that should not be present in the assembly. To determine if unigenes with >95% identity to reciprocal host sequences were due to cross contamination, the distribution of unigene pairwise identity was determined for a whole-plant normalized assembly (of an non-parasitic member of the Orobanchaceae, *Lindenbergia philippensis* (PPGP assembly LiPhGnB1, [23]). An identity threshold of 95% was established to minimize host contamination while retaining parasite transcripts with incidentally high identity to the host reference sequences.

Final assemblies were filtered by BLASTn to remove unigenes with >95% identity to host derived sequences (*Zea mays* and *Medicago truncatula,* transcripts from [38], mRNA ESTs from [39]). Remaining unigenes were then filtered by BLASTn to available PPGP transcriptome assemblies of*T. versicolor*[23] to select for unigenes with >95% identity to other putative *Triphysaria* transcripts. Unigenes with less than 95% identity to host or *T. versicolor* sequences were queried against the non-redundant protein sequences database [40] using BLASTx (1e-10). Sequences were sorted by best-hit species into non-plant and plant categories. Sequences that remained unannotated and unclassified after extensive efforts detailed above were translated based on the ESTScan ORF prediction and submitted to InterProScan [41] via blast2go [92] using default parameters. These unigenes were also submitted to the OrthoMCL DB [42] database using default parameters.

The unigenes with the greatest RPKM in unique and overlapping Orthogroups, excluding those belonging to Orthogroups shared in all three transcriptomes and those unique to the above ground transcriptome, were used to query the non-redundant protein sequences database (NR) at NCBI[40] with BLASTx (1e-10) using blast2go [92]. This set of 120 unigenes was queried against a collection of 17 annotated plant genomes using BLASTn and BLASTx to determine best hits to annotated plant sequences not necessarily present in NR. The 17 genomes queried are as follows:*Arabidopsis thaliana, Carica papaya, Fragaria vesca, Glycine max, Medicago truncatula, Mimulus guttatus, Oryza sativa, Phoenix dactylifera, Populus trichocarpa, Selaginella moellendorffii, Solanum lycopersicum, Solanum tuberosum, Sorghum bicolor, Thellungiella parvula, Theobroma cacao, Vitis vinifera,* and. Additionally, these unigenes were annotated with InterProScan [41] via blast2go [92] with the default settings.

GO Slim Category Analysis

An annotated *T. versicolor* PPGP transcriptome with no tissue overlap of interface transcriptomes (TrVeBC1, [23]) was included in Orthogroup analysis to serve as a point of comparison for interface transcriptome analysis. Putative *T. versicolor* unigenes (excluding host and non-plant unigenes) were sorted based on Orthogroup assignment using Venny [93]. GO Slim annotations from unigenes present in Orthogroups were subject to a Chi-Square test using R [94]. Alpha was set to 0.05.

Phylogenetic Analysis of Tvexpb1 and Tvexpa4

Homologs of TvEXPB1 from *Arabidopsis thaliana*, *Oryza sativa*, *Mimulus guttatus*, and *Selaginella moellendorfii* were extracted from Phytozome [38] v7.0: gene family #28891348 (β-expansins). Separately, alpha expansin homologs of TvEXPA4 were extracted from PlantTribes 2.0 [36] Orthogroup 1292. These sequences were combined with a subset of translatable sequences assembled in the PPGP project that had best BLAST hits with PlantTribes v2.0 Orthogroup 6163 (β-expansins) and 1292 (α-expansins). Inferred amino acid sequences for each data set were aligned using MUSCLE [95] and the coding DNA sequences were then forced onto this alignment for phylogenetic analysis. RAxML [96] version 7.2.8 was used to estimate the maximum likelihood tree under the GTR+gamma model of molecular evolution and bootstrap support values were estimated using 100 rapid bootstrapping replicates.

Verification of Expansin Transcript Sequence and Relative Transcript Abundance

cDNA Synthesis

cDNA was synthesized from the same amplified samples used for sequencing, plus an additional biological replicate of each, to use as a template for qRT-PCR using the iScript™ Reverse Transcription Supermix for RT-qPCR (BioRad). The manufacturer's instructions were followed for all RNA samples as this kit utilizes both oligo-dT and random priming.

RNA Isolation from Host Roots

RNA was isolated from host roots that were co-cultured, as described above, with *T. versicolor*. Host roots were harvested at sites adjacent to haustoria and flash frozen on liquid N_2. Approximately 50 mg of tissue was vigorously macerated in Kimble Chase glass tissue grinders (part # KT885450-0020) in the presence of 450 μL buffer RLT + β-mercapto-ethanol (Qiagen) for 3 min. The lysate was then transferred to a QiaShredder column and the RNeasy Plant Mini Kit (Qiagen) instructions were then followed with the addition of an on-column DNase treatment (Appendix D: RNeasy Mini Handbook 4th Edition) until elution, which was done in 30 μL RNase free water. Total RNA was assessed on the Agilent Bioanalyzer using the RNA 6000 Pico kit (Agilent) with the Plant Total RNA assay.

Primer Design and Sequence Verification

Primers were designed using Geneious Pro (v5.5.4 [97]) to amplify TvEXPA4 and TvEXPB1 from aRNA based upon the unigenes that resulted from the *de novo* assembly and post-processing of the Illumina paired-end mRNA Seq data.

Primer sequences:

TvEXPA_F5: GCTTTTGCCTACGACCAACTTATG

TvEXPA_R3: GACAGTTTTGCCATCGCTTGTAG

TvEXPB_F1: GCCATAGTTTCAACCCGAGGAC

TvEXPB_R2: GGCTTCTTCCTGCTCTCCTTACTTG

With these primers the putative α- and β-expansins transcripts were amplified and submitted for Sanger sequencing. Sequencing reads were quality trimmed manually by visually examining the electroferrograms. MUSCLE was used to align the Sanger reads with the unigenes to confirm the unigene sequence. Any remaining gaps were sequenced with the same method using primers based on the Sanger sequence verified first-round PCR products.

Quantitative Real-Time Polymerase Chain Reaction

(*qRT-PCR*) To verify the host-specific expression pattern of TvEXPB1 and TvEXPA4 observed in the *de novo* read mapping results, transcript levels of TvEXPB1, TvEXPA4, TvActin, ZmActin, and MtActin were estimated using qRT-PCR.

Primer sequences:

TvEXPA4:

For: 5′-TGGGAGGTGCTTGTGGGTAT-3′; Rev: 5′-CCGCAGGATAACCCATTGTT-3′

TvEXPB1:

For: 5′-GATGGCCTGACTGAAGTTGCA-3′; Rev: 5′-GCGGCAAATTCACCCTAAAA-3′

TvActin:

For: 5′-ACCCGATCCTTCTCACTGA-3′; Rev: 5′-CATGACAATACCAGTCGTACG-3′

ZmActinB:

For: 5′-CAATGGCACTGGAATGGT-3′; Rev: 5′-ATCTTCAGGCGAAACACG-3′ [98]

MtActin:

For: 5′-ATGTTGCTATTCAGGCCG-3′; Rev: 5′-GCTCATAGTCAAGGGCAAT-3′ [99]

Verification of Primer Specificity

To determine if the primers were specific to their intended targets, melt-curve analysis was performed for each PCR product. Primer specificity for parasite target genes was verified by submitting host RNA extracted from co-cultured host roots to analysis by qRT-PCR with parasite gene primers. In both cases host actin transcripts were detected, yet primer pairs specific to the parasite genes yielded signal consistent with background. All no-template controls (NTC) showed signal consistent with background signal and all reverse transcription negative (RTN) controls showed signal consistent with background.

qRT-PCR Assay Conditions

The qRT-PCR reaction prepared using the KAPA™ SYBR FAST qPCR kit (KAPA Biosystems, KK4602) following the manufacturer's instructions. The reaction was run on a BioRad MyiQ (170–9770) with the following program:

95°C for 8 min. (initial melt)

95°C for 0.5 min. (cycle melt)

60°C for 0.5 min. (cycle anneal/extend)

Repeat 40 cycles

Melt Curve: 0.5°C increments from 95°C – 25°C.

qRT-PCR Data Analysis

Crossing point (Ct) values for each of 3 technical replicates were used to calculate the average Ct value. The 2^(-ddCt) method was used to calculate the fold-change in expression in each sample relative to the control [100]. A one-tailed, two-sample *t*-test assuming unequal variance was performed using R [94]; alpha was set to 0.05.

AUTHORS' CONTRIBUTIONS

Conception and design of PPGP transcriptome (JHW,,CWD, MPT, JIY) and interface transcriptome study (LAH, CWD); seeds and *Triphysaria* growth and assays: JIY; plant cultivation and tissues: LAH; laser capture methods and sampling: LAH, CGT; RNAs, amplifications, and sequencing libraries: LAH; qRT-PCR: LAH, ZY; data analysis and presentation: LAH, JPD, EKW, NJW, ZY, NSA, CWD; Wrote manuscript: LAH and CWD with contributions from all of the authors. All authors read and approved the final manuscript.

ACKNOWLEDGMENTS

The authors gratefully acknowledge partial funding for this project through NSF Plant Genome award DBI-0701748 to Westwood JH, dePamphilis CW, Timko MP, and Yoder JI, NSF award IOS-0821954 and IOS-1104334 to Taylor CG, McIntyre L and Koch, K, and the Intercollege Graduate Program in Plant Biology and Department of Biology of Penn State University. We thank Paula Ralph for lab assistance and Yongde Bao (University of Virginia) for Illumina sequencing. We also thank Daniel J. Cosgrove for helpful discussion and encouragement throughout the course of this research. We thank dePamphilis lab members past and present for their assistance with various aspects of this work, and two anonymous reviewers for helpful comments.

REFERENCES

1. Westwood JH, Yoder JI, Timko MP, dePamphilis CW: The evolution of parasitism in plants. *Trends Plant Sci* 2010, 15(4):227-235.
2. Olmstead RG, dePamphilis CW, Wolfe AD, Young ND, Elisons WJ, Reeves PA:Disintegration of the Scrophulariaceae. *Am J Bot* 2001, 88(2):348-361
3. Ejeta G: The *Striga* scourge in Africa: a growing pandemic. In *Integrating New Technologies for Striga Control: Towards Ending the Witch-hunt*. Edited by Gressel J. Singapore: World Scientific Publishing Co; 2007:3-16.
4. Scholes JD, Press MC: *Striga* infestation of cereal crops-an unsolved problem in resource limited agriculture. *Curr Opin Plant Biol* 2008, 11(2):180-186
5. Parker C: Observations on the current status of *Orobanche* and *Striga* problems worldwide. *Pest Manag Sci* 2009, 65(5):453-459.
6. Vurro M, Bonciani B, Vannacci G: Emerging infectious diseases of crop plants in developing countries: impact on agriculture and socio-economic consequences. *Food Secur* 2010, 2(2):113-132.
7. Press MC, Graves JD: *Parasitic Plants*. London: Chapman and Hall; 1995.
8. Tomilov A, Tomilova N, Yoder JI: *Agrobacterium tumefaciens* and *Agrobacterium rhizogenes* transformed roots of the parasitic plant *Triphysaria versicolor* retain parasitic competence. *Planta* 2007, 225(5):1059-1071.
9. Yoder JI: A species-specific recognition system directs haustorium development in the parasitic plant *Triphysaria* (Scrophulariaceae). *Planta* 1997, 202(4):407-413
10. Westwood JH, de Pamphilis CW, Das M, Fernández-Aparicio M, Honaas LA, Timko MP, Wafula EK, Wickett NJ, Yoder JI: The parasitic plant genome project: new tools for understanding the biology of *Orobanche* and *Striga*. *Weed Sci* 2012, 60(2):295-306.
11. Kuijt J: *The Biology of Parasitic Flowering Plants*. Berkeley: University of California Press; 1969.

12. Heide-Jørgensen H, Kuijt J: Epidermal derivatives as xylem elements and transfer cells: a study of the host-parasite interface in two species of *Triphysaria*(Scrophulariaceae). *Protoplasma* 1993, 174(3):173-183.
13. Heide-Jørgensen HS, Kuijt J: The haustorium of the root parasite *Triphysaria*(Scrophulariaceae), with special reference to xylem bridge ultrastructure. *Am J Bot* 1995, 82(6):782-797.
14. Vachev T, Ivanova D, Minkov I, Tsagris M, Gozmanova M: Trafficking of the Potato spindle tuber viroid between tomato and *Orobanche ramosa*. *Virology* 2010, 399(2):187-193
15. Birschwilks M, Haupt S, Hofius D, Neumann S: Transfer of phloem-mobile substances from the host plants to the holoparasite *Cuscuta* sp. *J Exp Bot* 2006, 57(4):911-921
16. Tomilov AA, Tomilova NB, Wroblewski T, Michelmore R, Yoder JI: Trans-specific gene silencing between host and parasitic plants. *Plant J* 2008, 56(3):389-397
17. Thompson JN: *The Coevolutionary Process*. Chicago: The University of Chicago Press; 1994.
18. Agosta SJ, Janz N, Brooks DR: How specialists can be generalists: resolving the "parasite paradox" and implications for emerging infectious disease. *Zoologia (Curitiba)* 2010, 27:151-162.
19. Atsatt PR, Strong DR: Population biology of annual grassland hemiparasites .1. Host Environment. *Evolution* 1970, 24(2):278-291.
20. *The Parasitic Plant Connection*. http://www.parasiticplants.siu.edu/ *webcite*
21. Schneeweiss GM: Correlated evolution of life history and host range in the nonphotosynthetic parasitic flowering plants *Orobanche* and *Phelipanche*(Orobanchaceae). *J Evol Biol* 2007, 20(2):471-478.
22. Thurman LD: *Genecological studies in Orthocarpus subgenus Triphysaria*. University of California, Berkeley: PhD Thesis. Berkeley; 1966.
23. *The Parasitic Plant Genome Project*. http://ppgp.huck.psu.edu/ *webcite*

24. Estabrook EM, Yoder JI: Plant-plant communications: Rhizosphere signaling between parasitic angiosperms and their hosts. *Plant Physiol* 1998, 116(1):1-7.

25. Jamison DS, Yoder JI: Heritable variation in quinone-induced haustorium development in the parasitic plant *Triphysaria*. *Plant Physiol* 2001, 125(4):1870-1879.

26. Bandaranayake PCG, Filappova T, Tomilov A, Tomilova NB, Jamison-McClung D, Ngo Q, Inoue K, Yoder JI: A single-electron reducing quinone oxidoreductase is necessary to induce haustorium development in the root parasitic plant *Triphysaria*. *Plant Cell* 2010, 22(4):1404-1419.

27. Offler CE, McCurdy DW, Patrick JW, Talbot MJ: Transfer cells: Cells specialized for a special purpose. *Annu Rev Plant Biol* 2003, 54:431-454.

28. Wall PK, Leebens-Mack J, Chanderbali AS, Barakat A, Wolcott E, Liang H, Landherr L, Tomsho LP, Hu Y, Carlson JE: Comparison of next generation sequencing technologies for transcriptome characterization. *BMC Genom* 2009, 10(1):347.

29. Strickler SR, Bombarely A, Mueller LA: Designing a transcriptome next-generation sequencing project for a nonmodel plant species. *Am J Bot* 2012, 99(2):257-266.

30. Grabherr MG, Haas BJ, Yassour M, Levin JZ, Thompson DA, Amit I, Adiconis X, Fan L, Raychowdhury R, Zeng Q: Full-length transcriptome assembly from RNA-Seq data without a reference genome. *Nat Biotech* 2011, 29(7):644-652.

31. O'Malley RC, Lynn DG: Expansin message regulation in parasitic angiosperms: marking time in development. *Plant Cell* 2000, 12(8):1455-1465.

32. Wrobel RL, Yoder JI: Differential RNA expression of alpha-expansin gene family members in the parasitic angiosperm *Triphysaria versicolor* (Scrophulariaceae). *Gene* 2001, 266(1–2):85-93.

33. Kenzelmann M, Klaren R, Hergenhahn M, Bonrouhi M, Grone HJ, Schmid W, Schutz G: High-accuracy amplification of nanogram total RNA amounts for gene profiling. *Genomics* 2004, 83(4):550-558.

34. Matas AJ, Yeats TH, Buda GJ, Zheng Y, Chatterjee S, Tohge T, Ponnala L, Adato A, Aharoni A, Stark R: Tissue- and cell-type specific transcriptome profiling of expanding tomato fruit provides insights into metabolic and regulatory specialization and cuticle formation. *Plant Cell* 2011, 23(11):3893-3910.

35. Wall PK, Leebens Mack J, Muller KF, Field D, Altman NS, dePamphilis CW: PlantTribes: a gene and gene family resource for comparative genomics in plants. *Nucleic Acids Res* 2008, 36:D970-D976.

36. *The Floral Genome Project*. http://fgp.bio.psu.edu

37. Wickett NJ, Honaas LA, Wafula EK, Das M, Huang K, Wu B, Landherr L, Timko MP, Yoder J, Westwood JH: Transcriptomes of the parasitic plant family Orobanchaceae reveal surprising conservation of chlorophyll synthesis. *Curr Biol* 2011, 21:2098-2104. |

38. *Phytozome*. http://www.phytozome.net

39. *PlantGBD*. http://www.plantgdb.org

40. *National Center for Biotechnology Information*. http://www.ncbi.nlm.nih.gov

41. *European Bioinformatics Institute*. http://www.ebi.ac.uk/

42. *OrthoMCL DB*.http://www.orthomcl.org/

43. Sampedro J, Cosgrove DJ: The expansin superfamily. *Genome Biol* 2005, 6(12):242.

44. Brandt SP: Microgenomics: Gene expression analysis at the tissue-specific and single-cell levels. *J Exp Bot* 2005, 56(412):495-505.

45. Day RC, Grossniklaus U, Macknight RC: Be more specific! Laser-assisted microdissection of plant cells. *Trends Plant Sci* 2005, 10(8):397-406.

46. Nelson T, Gandotra N, Tausta SL: Plant cell types: Reporting and sampling with new technologies. *Curr Opin Plant Biol* 2008, 11(5):567-573.

47. Gomez SK, Harrison MJ: Laser microdissection and its application to analyze gene expression in arbuscular mycorrhizal symbiosis. *Pest Manag Sci* 2009, 65(5):504-511.

48. Woll K: Isolation, characterization, and pericycle-specific transcriptome analyses of the novel maize lateral and seminal

root initiation mutant rum1. *Plant Physiol* 2005, 139(3):1255-1267.

49. Ithal N, Recknor J, Nettleton D, Hearne L, Maier T, Baum TJ, Mitchum MG: Parallel genome-wide expression profiling of host and pathogen during soybean cyst nematode infection of soybean. *MPMI* 2007, 20(3):293-305.
50. Gomez SK, Javot H, Deewatthanawong P, Torres-Jerez I, Tang Y, Blancaflor EB, Udvardi MK, Harrison MJ: *Medicago truncatula* and *Glomus intraradices* gene expression in cortical cells harboring arbuscules in the arbuscular mycorrhizal symbiosis. *BMC Plant Biol* 2009, 9(1):10.
51. Klink VP, Hosseini P, Matsye P, Alkharouf NW, Matthews BF: A gene expression analysis of syncytia laser microdissected from the roots of the *Glycine max* (soybean) genotype PI 548402 (Peking) undergoing a resistant reaction after infection by*Heterodera glycines* (soybean cyst nematode). *Plant Mol Biol* 2009, 71(6):525-567.
52. Matas AJ, Agustí J, Tadeo FR, Talón M, Rose JKC: Tissue-specific transcriptome profiling of the citrus fruit epidermis and subepidermis using laser capture microdissection. *J Exp Bot* 2010, 61(12):3321-3330.
53. Pillitteri LJ, Peterson KM, Horst RJ, Torii KU: Molecular profiling of stomatal meristemoids reveals new component of asymmetric cell division and commonalities among stem cell populations in Arabidopsis. *Plant Cell* 2011, 23(9):3260-3275.
54. Emrich SJ, Barbazuk WB, Li L, Schnable PS: Gene discovery and annotation using LCM-454 transcriptome sequencing. *Genome Res* 2007, 17(1):69-73.
55. Ohtsu K, Smith MB, Emrich SJ, Borsuk LA, Zhou RL, Chen TL, Zhang XL, Timmermans MCP, Beck J, Buckner B: Global gene expression analysis of the shoot apical meristem of maize (*Zea mays L.*). *Plant J* 2007, 52(3):391-404.
56. Li PH, Ponnala L, Gandotra N, Wang L, Si YQ, Tausta SL, Kebrom TH, Provart N, Patel R, Myers CR: The developmental dynamics of the maize leaf transcriptome. *Nat Genet* 2010, 42(12):1060-1067.

57. Feldman AL, Costouros NG, Wang E, Qian M, Marincola FM, Alexander HR, Libutti SK:Advantages of mRNA amplification for microarray analysis. *Biotechniques* 2002, 33(4):906.
58. Polacek DC, Passerini AG, Shi C, Francesco NM, Manduchi E, Grant GR, Powell S, Bischof H, Winkler H, Stoeckert CJ: Fidelity and enhanced sensitivity of differential transcription profiles following linear amplification of nanogram amounts of endothelial mRNA. *Physiol Genomics* 2003, 13(2):147-156.
59. Li Y, Li T, Liu SZ, Qiu MY, Han ZY, Jiang ZL, Li RY, Ying K, Xie Y, Mao YM: Systematic comparison of the fidelity of aRNA, mRNA and T-RNA on gene expression profiling using cDNA microarray. *J Biotech* 2004, 107(1):19-28.
60. King C, Guo N, Frampton GM, Gerry NP, Lenburg ME, Rosenberg CL: Reliability and reproducibility of gene expression measurements using amplified RNA from laser-microdissected primary breast tissue with oligonucleotide arrays. *J Mol Diagnos* 2005, 7(1):57.
61. Day RC, McNoe LA, Macknight RC: Transcript analysis of laser microdissected plant cells. *Physiol Plantarum* 2007, 129(2):267-282.
62. Goldsworthy SM, Stockton PS, Trempus CS, Foley JF, Maronpot RR: Effects of fixation on RNA extraction and amplification from laser capture microdissected tissue. *Mol Carcino* 1999, 25(2):86-91.
63. Gillespie JW, Best CJM, Bichsel VE, Cole KA, Greenhut SF, Hewitt SM, Ahram M, Gathright YB, Merino MJ, Strausberg RL: Evaluation of non-formalin tissue fixation for molecular profiling studies. *Am J Path* 2002, 160(2):449-457.
64. Kerk NM: Laser capture microdissection of cells from plant tissues. *Plant Physiol* 2003, 132(1):27-35.
65. Nakazono M: Laser-capture microdissection, a tool for the global analysis of gene expression in specific plant cell types: identification of genes expressed differentially in epidermal cells or vascular tissues of maize. *Plant Cell* 2003, 15(3):583-596.
66. Rispail N, Dita MA, Gonzalez-Verdejo C, Perez-de-Luque A, Castillejo MA, Prats E, Roman B, Jorrin J, Rubiales D: Plant resistance to parasitic plants: molecular approaches to an old foe. *New Phytol* 2007, 173(4):703-711.

67. Tomilov A, Tomilova N, Shin DH, Jamison D, Torres J, Reagan R, Horning T, Truong R, Nava A, Nava A: Chemical signaling between plants: mechanistic similarities between phytotoxic allelopathy and host recognition by parasitic plants. In *Chemical Communication: From Gene to Ecosystem*. Edited by Dicke M. Wageningen: Frontis; 2006:55-69.
68. Cosgrove DJ: Growth of the plant cell wall. *Nat Rev Mol Cell Biol* 2005, 6(11):850-861.
69. Albrecht U, Bowman KD: Gene expression in *Citrus sinensis* (L.) Osbeck following infection with the bacterial pathogen *Candidatus liberibacter* asiaticus causing Huanglongbing in Florida. *Plant Sci* 2008, 175(3):291-306.
70. Ding XH, Cao YL, Huang LL, Zhao J, Xu CG, Li XH, Wang SP: Activation of the indole-3-acetic acid-amido synthetase GH3-8 suppresses expansin expression and promotes salicylate- and jasmonate-independent basal immunity in rice. *Plant Cell* 2008, 20(1):228-240.
71. Gal TZ, Aussenberg ER, Burdman S, Kapulnik Y, Koltai H: Expression of a plant expansin is involved in the establishment of root knot nematode parasitism in tomato. *Planta* 2006, 224(1):155-162.
72. Kikuchi T, Aikawa T, Kosaka H, Pritchard L, Ogura N, Jones JT: Expressed sequence tag (EST) analysis of the pine wood nematode *Bursaphelenchus xylophilus* and *B-mucronatus*. *Mol Biochem Parasitol* 2007, 155(1):9-17.
73. Li ZC, Durachko DM, Cosgrove DJ: An oat coleoptile wall protein that induces wall extension in-vitro and that is antigenically related to a similar protein from cucumber hypocotyls. *Planta* 1993, 191(3):349-356.
74. Cho H-T, Kende H: Expansins in deepwater rice internodes. *Plant Physiol* 1997, 113(4):1137-1143.
75. Cosgrove DJ, Bedinger P, Durachko DM: Group I allergens of grass pollen as cell wall-loosening agents. *P Natl Acad Sci USA* 1997, 94(12):6559.
76. McQueenmason SJ, Cosgrove DJ: Disruption of hydrogen bonding between plant cell wall polymers by proteins that induce wall extension. *P Natl Acad Sci USA* 1994, 91:6574-6578.

77. McQueenmason SJ, Cosgrove DJ: Expansin mode of action on cell-walls - Analysis of wall hydrolysis, stress-relaxation, and binding. *Plant Physiol* 1995, 107(1):87-100.

78. Li Y, Jones L, McQueen-Mason S: Expansins and cell growth. *Curr Opin Plant Biol* 2003, 6(6):603-610.

79. Cosgrove DJ: Loosening of plant cell walls by expansins. *Nature* 2000, 407(6802):321-326.

80. Kapu NUS, Cosgrove DJ: Changes in growth and cell wall extensibility of maize silks following pollination. *J Exp Bot* 2010, 61(14):4097-4107.

81. Tabuchi A, Li LC, Cosgrove DJ: Matrix solubilization and cell wall weakening by beta-expansin (group-1 allergen) from maize pollen. *Plant J* 2011, 68(3):546-559.

82. Yennawar NH, Li L-C, Dudzinski DM, Tabuchi A, Cosgrove DJ: Crystal structure and activities of EXPB1 (Zea m 1), alpha, beta-expansin and group-1 pollen allergen from maize. *P Natl Acad Sci USA* 2006, 103(40):14664-14671.

83. Vial L, Chapalain A, Groleau MC, Deziel E: The various lifestyles of the *Burkholderia cepacia* complex species: a tribute to adaptation. *Environ Microbiol* 2011, 13(1):1-12.

84. Atsatt PR: Parasitic flowering plants: How did they evolve? *Am Nat* 1973, 107(956):502-510.

85. Chuang TI, Heckard LR: Generic realignment and synopsis of subtribe Castillejinae (Scrophulariaceae-tribe Pedicularieae). *Syst Bot* 1991, 16(4):644-666.

86. Fernandez-Aparicio M, Rubiales D, Bandaranayake P, Yoder J, Westwood J:Transformation and regeneration of the holoparasitic plant *Phelipanche aegyptiaca*. *Plant Methods* 2011, 7(1):36.

87. Torres MJ, Tomilov AA, Tomilova N, Reagan RL, Yoder JI: Pscroph, a parasitic plant EST database enriched for parasite associated transcripts. *BMC Plant Biol* 2005, 5:24.

88. Schroeder A, Mueller O, Stocker S, Salowsky R, Leiber M, Gassmann M, Lightfoot S, Menzel W, Granzow M, Ragg T: The RIN: an RNA integrity number for assigning integrity values to RNA measurements. *BMC Mol Biol* 2006, 7:3.

89. *The MarthLab.* http://bioinformatics.bc.edu/marthlab/Main_Page

90. Iseli C, Jongeneel CV, Bucher P: ESTScan: a program for detecting, evaluating, and reconstructing potential coding regions in EST sequences. *Proc Int Conf Intell Syst Mol Biol* 1999, 138-148.
91. Edgar RC: Search and clustering orders of magnitude faster than BLAST. *Bioinformatics* 2010, 26(19):2460-2461
92. *blast2go*. http://www.blast2go.com/
93. *Venny*. http://bioinfogp.cnb.csic.es/tools/venny/index.htm
94. *The R Project for Statistical Computing*. http://www.r-project.org/
95. Edgar RC: MUSCLE: multiple sequence alignment with high accuracy and high throughput. *Nucleic Acids Res* 2004, 32:1792-1797.
96. Stamatakis A: RAxML-VI-HPC: Maximum Likelihood-based Phylogenetic Analyses with Thousands of Taxa and Mixed Models. *Bioinformatics* 2006, 22:2688-2690.
97. *Geneious*. http://www.geneious.com/
98. Xing GF, Guo GG, Yao YY, Peng HR, Sun QX, Ni ZF: Identification and characterization of a novel hybrid upregulated long non-protein coding RNA in maize seedling roots. *Plant Sci* 2010, 179(4):356-363.
99. Jones KM, Sharopova N, Lohar DP, Zhang JQ, VandenBosch KA, Walker GC: Differential response of the plant Medicago truncatula to its symbiont *Sinorhizobium meliloti* or an exopolysaccharide-deficient mutant. *P Natl Acad Sci USA* 2008, 105(2):704-709.
100. Pfaffl MW: A new mathematical model for relative quantification in real-timeRT-PCR. *Nucleic Acids Res* 2001, 29(9):e45.

Chapter 3

Functional Analysis of the Theobroma Cacao NPR1 Gene in Arabidopsis

Zi Shi[1], Siela N Maximova[2], Yi Liu[1], Joseph Verica[2], and Mark J Guiltinan[1,2]

[1]Huck Institute of Life Sciences, The Pennsylvania State University, University Park, PA 16802, USA

[2]The Department of Horticulture, The Pennsylvania State University, University Park, PA 16802, USA

ABSTRACT

Background

The *Arabidopsis thaliana NPR1* gene encodes a transcription coactivator (NPR1) that plays a major role in the mechanisms regulating plant defense response. After pathogen infection and in response to salicylic acid (SA) accumulation, NPR1 translocates from the cytoplasm into the nucleus where it interacts with other transcription factors resulting in increased expression of over 2000 plant defense genes contributing to a pathogen resistance response.

Results

A putative *Theobroma cacao NPR1* cDNA was isolated by RT-PCR using degenerate primers based on homologous sequences from *Brassica, Arabidopsis* and *Carica papaya*. The cDNA was used to isolate a genomic clone from *Theobroma cacao* containing a putative *TcNPR1* gene. DNA sequencing revealed the presence of a 4.5 kb coding region containing three introns and encoding a polypeptide of 591 amino acids. The predicted TcNPR1 protein shares 55% identity and 78% similarity to *Arabidopsis* NPR1, and contains each of the highly conserved functional domains indicative of this class of transcription factors (BTB/POZ and ankyrin repeat protein-protein interaction domains and a nuclear localization sequence (NLS)). To functionally define the *TcNPR1*gene, we transferred *TcNPR1* into an *Arabidopsis npr1* mutant that is highly susceptible to infection by the plant pathogen *Pseudomonas syringae* pv. tomato DC3000. Driven by the constitutive CaMV35S promoter, the cacao *TcNPR1* gene partially complemented the *npr1* mutation in transgenic *Arabidopsis* plants, resulting in 100 fold less bacterial growth in a leaf infection assay. Upon induction with SA, *TcNPR1* was shown to translocate into the nucleus of leaf and root cells in a manner identical to *Arabidopsis* NPR1. Cacao NPR1 was also capable of participating in SA-JA signaling crosstalk, as evidenced by the suppression of JA responsive gene expression in *TcNPR1*overexpressing transgenic plants.

Conclusions

Our data indicate that the *TcNPR1* is a functional ortholog of *Arabidopsis NPR1*, and is likely to play a major role in defense response in cacao. This fundamental knowledge can contribute to breeding of disease resistant cacao varieties through the application of molecular markers or the use of transgenic strategies.

BACKGROUND

Plants have evolved a complex network of defense responses, often associated with a response local to the site of infection [1-4]. In addition, defenses are also systemically induced in remote parts of the plant in a process known as systemic acquired resistance (SAR) [2,5,6]. Induction of the SAR pathway leads to heightened broad-spectrum resistance to secondary pathogen attacks by a variety of pathogens. Multiple studies in both monocots and dicots have shown that salicylic acid (SA) plays a central role as a signaling molecule in SAR [7-14]. Following pathogen attack, SA levels increase both locally and systemically in infected plants. In addition, SA is required for the induced expression of a set of pathogenesis-related (*PR*) genes [7,15-17].

NPR1 was originally identified by screening for mutants that were insensitive to SA (or its chemical analogs, 2,6-dichloroisonicotic acid (INA) or benzothiadiazole (BTH)) in *Arabidopsis* [7,18-20]. These screens identified a mutation designated as *Non-Expressor of PR*

1 (*NPR1*). Studies that followed further documented that *npr1* mutants displayed reduced expression of *PR* genes upon SA treatment and were more susceptible to pathogens [7,18,20,21]. Conversely, when *NPR1* was overexpressed, the resulting transgenic plants displayed increased resistance to pathogens, and were able to induce increased levels of *PR* genes in a dose-dependent fashion [22].

NPR1 encodes a protein containing ankyrin repeats and a BTB/POZ domain, both of which mediate protein-protein interactions in animals [23]. NPR1 shares homology with IκBα transcription inhibitors, which regulate the innate immunity response [21,24]. Recent work has shed light onto the mechanisms of NPR1 function [5,6,10,17,25-27]. *NPR1* is constitutively expressed, and NPR1 protein is present as inactive oligomers in the cytoplasm of the cell. Upon SAR induction, the redox state of the cell is altered, resulting in the reduction of NPR1 to its active monomeric form. Monomeric NPR1 moves into the nucleus where it can affect the induction of *PR* genes. Although NPR1 itself has no DNA binding domains, it participates in the regulation of defense gene transcription via interactions with TGA transcription factors [16,28-33]. In *Arabidopsis*, two conserved cysteine residues (C82 and C216) have been shown to be essential to the oligomerization and cytoplasmic localization of AtNPR1 [25]. Mutation of these residues results in constitutive monomerization and nuclear localization of NPR1.

It is believed that NPR1 also plays a role in the jasmonic acid (JA) signaling pathway and mediates the crosstalk between SA-JA defense pathways to fine-tune defense responses [27,30,34-36]. SA-mediated defenses are mainly effective against biotrophic pathogens, whereas JA-mediated defenses are predominantly efficient against necrotrophic pathogens and herbivorous insects. NPR1 mediates the antagonistic effect of SA on JA signaling by suppressing the expression of JA-responsive genes upon combined treatment of SA and methyl jasmonate (MeJA) [34].

A growing body of evidence has revealed that the salicylic acid dependent, NPR1-mediated defense pathway is also conserved in other plant species across wide phylogenetic distances. Two*NPR1-like* genes have been characterized from *Vitis vinifera* (grapevine) [14]. When translational fusions of the proteins encoded by the two genes with GFP were transiently expressed in *Nicotiana benthamiana* leaves, the proteins were localized predominantly to the nucleus and triggered the accumulation of pathogenesis-related proteins PR1 and PR2. In addition, the silencing of a tomato*NPR1-like* gene leads to increased bacterial growth upon *Ralstonia solanacearum* infection in tomato [12]. In tobacco, the suppression of *NPR1-like* gene leads to increased susceptibility to tobacco mosaic virus [8]. Similarly, overexpression of the apple *MpNPR1* gene in transgenic apple plants resulted in the up-regulation of *PR* genes and enhanced resistance to bacterial and fungal pathogens [37]. In wheat, the expression of *Arabidopsis NPR1* confers resistance to Fusarium head blight in susceptible cultivar Bobwhite [13]. Major efforts have been made to study the SA and NPR1-dependent pathway in rice, the model monocot plant. Treatment of rice plants with the salicylic acid analog probenazole results in enhanced resistance against rice blast fungus [38]. In addition, rice plants expressing bacterial salicylate hydrolase (*nahG*) are unable to accumulate salicylic acid and display increased susceptibility to rice blast [39]. Overexpression of the*Arabidopsis NPR1* gene in rice leads to enhanced resistance to the bacterial pathogen*Xanthomonas oryzae* pv. *oryzae* [9]. An orthologue of *NPR1* has been isolated from rice (*OsNPR1/NH1*), and the overexpression of *OsNPR1* in rice leads to enhanced resistance to both bacterial and oomycete pathogens [40]. Moreover, *OsNPR1* is able to complement the *Arabidopsis npr1-1* mutant [11]. Like AtNPR1, OsNPR1 is also constitutively expressed and localizes to the cytoplasm. Treatment of rice cells with a reducing agent resulted in the movement of OsNPR1 into the nucleus. Similar to *Arabidopsis* NPR1, mutation of the corresponding cysteines (C82 and C216) in OsNPR1 also resulted in constitutive nuclear localization [11]. Thus, it appears that the mechanisms of SA-dependent, NPR1-mediated defense response likely evolved very early in the emergence of the plant kingdom.

Theobroma cacao L, (cacao) is a small tropical tree species endemic to the Amazon rainforest of South America. Cacao seeds are harvested and processed into cocoa beans and chocolate, providing an income for millions of small-holder farmers in West Africa, Central and South America, the Caribbean, Malaysia, Indonesia and other tropical areas. Pathogens are a major problem for cacao production, causing annual crop losses estimated at 30-40% [41]. In its center of diversity, the Amazon basin, cacao is susceptible to several potentially devastating pathogens, such as *Moniliophthora perniciosa*, the causal agent of witches' broom disease, *Moniliophthora roreri*, the causal agent of frosty pod rot [41-45] and several *Phytophthora* spp., the causal agent of black pod disease [46,47]. Outside this region, cacao is susceptible to a number of opportunistic pathogens [48-50].

Several defense-related genes in *Theobroma cacao* have been identified through gene expression analyses after hormone treatments [46,47,51]. An endo-1,4-β-glucanase is induced by the application of ethylene, and a type III peroxidase and a class VII chitinase are induced by methyl jasmonate treatment in mature cacao leaves. Those genes are responsible for induced resistance to pests in cacao, though the responses to hormone induction are different depending on developmental stages. In addition, transgenic overexpression of a class I chitinase gene in cacao enhances foliar resistance against the fungal pathogen, *Colletotrichum gloeosporioides* [52]. Moreover, ESTs sharing sequence homology to known *PR* genes have been isolated from cacao[53-55]. Several of these genes have been shown to be up-regulated by treatment of plants with benzothiadiazole (BTH), the salicylic acid analog [53]. All together, recent evidence suggests that cacao may utilize SAR pathway during the defense response; however, the extent of conservation of the pathway in cacao is presently unknown.

In this paper, we report the isolation and characterization of an *NPR1* homologue from the tropical tree, *Theobroma cacao*. We show that *Theobroma cacao NPR1* (*TcNPR1*) shares similar functions as *Arabidopsis NPR1*. It is able to partially complement the *Arabidopsis npr1-2* mutation in transgenic *Arabidopsis* plants in a leaf infection assay and translocate into nucleus upon SA induction in the same manner as the endogenous *Arabidopsis* NPR1 protein.

RESULTS

Isolation of a Putative *TcNPR1* Gene

Degenerate PCR was utilized to clone the full length cDNA of *Theobroma cacao NPR1* (*TcNPR1*). The degenerate primers were designed based on the alignment of NPR1 homologs from*Arabidopsis*, *Brassica* and *Carica papaya* and cDNA from cacao genotype Scavina6 (SCA6) leaf was used as template. A fragment of 1776 bp was isolated, cloned into pGEM sequencing vector and sequenced to reveal an intact coding sequence of the expected length and with high homology to the *Arabidopsis NPR1* gene.

A genomic fragment containing a putative *TcNPR1* gene was obtained by screening Clemson University Genomics Institute (CUGI) cacao BAC library using the putative cacao *TcNPR1* cDNA clone as probe. Two BAC clones were found to contain the *TcNPR1* gene: 2K13 and 11K17. The genomic sequence of *TcNPR1* was isolated by primer walking sequencing from known sequence using clone 2K13. A similar strategy was performed to sequence a region of 1.1 kb containing the promoter sequence upstream of ATG start codon. The full sequence consisted of a 4.5 kb genomic region of *TcNPR1* containing 1.1 kb promoter, four exons and three introns (depicted in Figure 1A), which is similar to the genomic structure of *AtNPR1*.

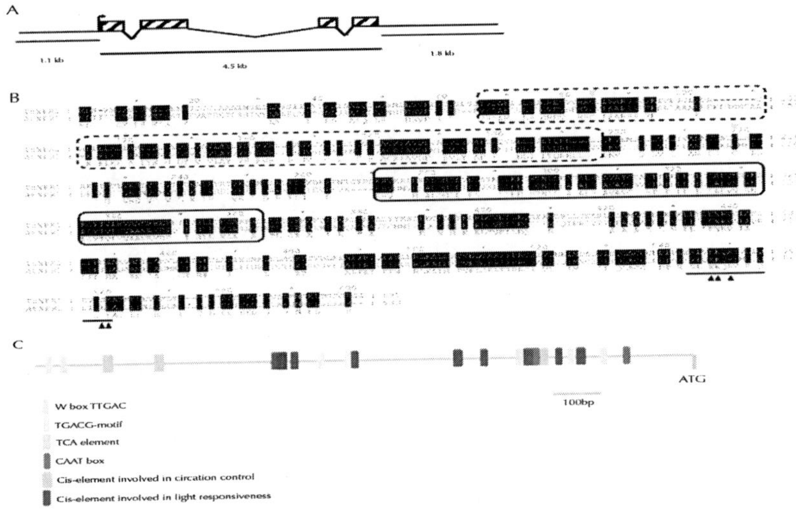

Figure 1: Gene and protein structures of *Theobroma cacao NPR1*. A. Diagram of *TcNPR1* gene structure. Boxes with diagonal stripes represent exons. Diagonal lines represent introns. The arrow represents the start site of transcription. The sizes of the promoter region, coding and the 3′-untranslated (UTR) regions of *TcNPR1* are indicated. B. Alignment of AtNPR1 and TcNPR1 proteins. Protein alignment was carried out by ClustalW. Residues blocked in black are identical in both sequences. Numbers refer to the amino acid position in AtNPR1 protein. BTB/POZ and ankyrin repeats domains are highlighted by dashed line box and solid line box, respectively. Two of the conserved cysteines (C82 and C216 in AtNPR1) are shown with grey triangles. The potential nuclear localization signal identified in *Arabidopsis* is underlined. Amino acids demonstrated to be critical for AtNPR1 nuclear translocation are indicated with black triangles. C. Schematic representation of predicted cis-acting regulatory DNA element in cacao *TcNPR1* promoter region. A 1.1 kb DNA fragment upstream of start codon was analyzed by querying the PLACE and PlantCare databases. The colored blocks represent different cis-elements as indicated.

Arabidopsis and Cacao NPR1 Protein Sequences are Highly Similar

Conceptual translation of the cacao NPR1 protein revealed that it consists of 591 amino acid residues, only two amino acids shorter than

AtNPR1. Alignment of the AtNPR1 and TcNPR1 protein sequences revealed that they are highly similar to each other (55% identity, 74% similarity). Both the *Arabidopsis* and cacao *NPR1* genes encode predicted proteins that share a number of structural features (Figure 1B). Each has a BTB/POZ domain near its N-terminal end (dashed line box) which shares 65% identity. Similarly, an ankyrin repeat region (solid line box) is present in both proteins which shares about 72% identity. In other ankyrin containing proteins, these domains have been shown to play roles in protein-protein interactions [16,23,56,57]. In the AtNPR1 protein, the BTB/POZ domain has been shown to function in homo-dimerization of NPR1, and the ankyrin repeat region mediates interactions with TGA transcription factors [58]. In addition, two cysteine residues (C82 and C216 in AtNPR1), which have been shown to play a role in the redox regulated activation and nuclear localization [25], are also conserved in TcNPR1 (Figure 1B. grey triangles). In fact, the AtNPR1 and TcNPR1 proteins share eleven conserved cysteine residues, suggesting that they share a similar structural conformation. The C-terminal region of AtNPR1 has been shown to contain a nuclear localization signal (NLS) that directs NPR1 monomers into the nucleus upon induction [59]. Five basic amino acids in this region function directly in this role (Figure 1B, black arrows). Four out of five of these basic amino acids are identical in TcNPR1, suggesting that TcNPR1 may also contain functional nuclear localization sequences. These similarities in protein structure suggest that *TcNPR1* gene may also share the same function as *AtNPR1* during plant defense response.

Cacao NPR1 Gene Promoter Contains Putative SA Regulatory Elements

We analyzed the 1.1 kb promoter region of the TcNPR1 gene (Figure 1C) using plant cis-acting regulatory elements databases PLACE http://www.dna.affrc.go.jp/PLACE/ [60] and PlantCare http://bioinformatics.psb.ugent.be/webtools/plantcare/html/ [61,62]. Although a potential CAAT box was found 290 bp and 140 bp upstream of the ATG start codon, we did not observe an element resembling a TATA box. This is not surprising, as recent studies of core promoter regions in both plants and animals suggest that only 24%-29% of genes contain TATA-like elements [63,64]. A variety of other regulatory elements were also found. Several elements known to regulate inducibility by salicylic acid were found, such as the AS-1 element (TGACG). TGACG motifs were found involved in transcription activation by SA and this element was previously shown to be required for the SA-induced expression of *PR1* [65]. In addition, there were multiple copies of the W-box (TTGAC), an element similar to the AS-1 element, which was also found in promoter of *AtNPR1*. W-box was shown to be the binding site for SA-induced WRKY DNA binding proteins [66], and was required for the SA induction of the tobacco (*Nicotiana tabacum*) class I chitinase gene [67]. All of the information suggests that the *TcNPR1* gene might be regulated by SA in a manner similar to *AtNPR1*. Interestingly, several cis-elements involved in light responsiveness and circadian control are also presented in the *TcNPR1* promoter, suggesting that *TcNPR1* might be also regulated by light.

Basal and Induced Expression of *TcNPR1* in Cacao Tissues

Semi-quantitative RT-PCR was performed to illustrate the basal expression level of *TcNPR1* in various cacao tissues of Scavina6, including leaves from stage A (young/expanding), C (expanded/soft), E (mature/hardened), open flowers, unopened flowers, roots, seeds and fruit exocarps. *TcNPR1* transcript was detected in all tissues tested (Figure 2A), an expression pattern similar to the *Arabidopsis* gene, however, the basal level of expression varies among different tissues. The expression of *TcNPR1* was relatively high in the younger leaves (stage A and C) and lower in the later stages of development (stage E). The lowest expression of *TcNPR1* in all tested tissue was observed in seeds whereas the expression was relatively high in fruit exocarps. In flowers, expression of *TcNPR1* was higher in open flowers than in unopened ones. The expression of *TcNPR1* in roots was at a moderate level, comparable to that in flowers and younger leaves.

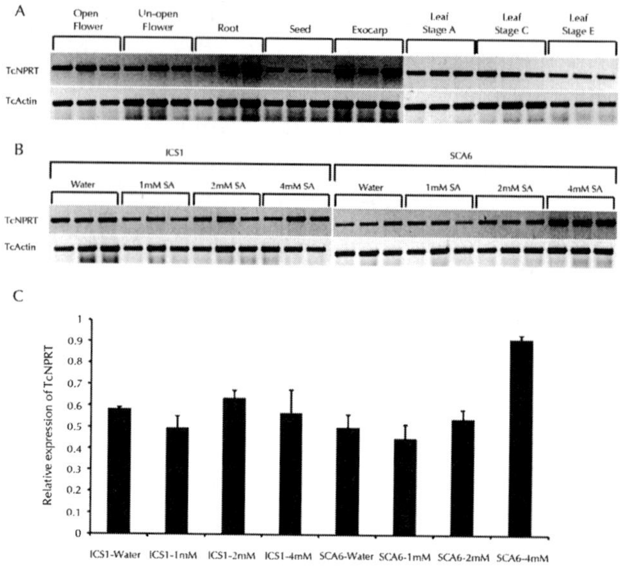

Figure 2: Gene expression analysis of *TcNPR1* in cacao. A. Expression of *Tc-NPR1* in various cacao tissues. Total RNA samples were collected from open flowers, unopened flowers, roots, seeds, exocarp and three different leaf developmental stages from youngest to oldest (A, C and E) from cacao genotype Scavina6 (SCA6). Semi- quantitative RT-PCR was performed and cacao *actin* (*TcActin*) was used as cDNA loading control. B. Expression of *TcNPR1* in cacao leaf tissue after salicylic acid (SA) treatment. Semi-quantitative RT-PCR was performed with cDNA from stage C leaves of two different cacao genotypes ICS1 (left panel) and SCA6 (right panel), sampled 24 hrs after SA treatment in three different concentrations (1 mM, 2 mM and 4 mM). Water-treated samples served as a control and *TcActin* was used as cDNA normalization control. C. Calculated average relative gene expression levels from B. Gel images were quantified by ImageQuant and expression of*TcNPR1* was normalized to *TcActin*. Expression levels are presented as the means ± standard errors of three biological replicates.

Induction by SA

Since it is well-characterized that *NPR1* transcript accumulation can be increased by SA treatment of *Arabidopsis* leaves, we tested if *TcNPR1* can respond to exogenous SA in the same manner. We applied various concentrations of SA to stage C leaves of two genotypes, Scavina6 and ICS1, which differ in their resistance to witches' broom disease (Scavina6 is more resistant) [68]. In*Arabidopsis*, the *NPR1* gene is induced approximately 2-3 fold 24 hrs after treatment of leaves with 1 mM SA [2,69]. Semi-quantitative RT-PCR was employed to demonstrate the induced level of*TcNPR1* 24 hours after SA application (Figure 2B). To quantify the expression of *TcNPR1* after SA treatment, we measured the fluorescence intensity of ethidium bromide stained DNA fragments irradiated with UV light using a high-sensitivity camera and ImageQuant software. Data were normalized to the expression level of an *actin* control. The results presented in Figure 2B showed that there was no significant change of *TcNPR1* expression upon 1 mM, 2 mM and 4 mM SA treatment in ICS1 compared to water control. However, in the Scavina6, there was a statistically-significant 2-fold increase of *TcNPR1* at 4 mM SA induction, though there was no change upon 1 mM and 2 mM SA treatment.

Complementation of *Arabidopsis Npr1-2* Mutant

To assess the function of TcNPR1, we placed the cacao *TcNPR1* gene under the control of the E-12 omega promoter and introduced it into the *Arabidopsis npr1-2* mutant to test if it can restore the mutant phenotype. One of the well characterized phenotypes of this mutant is the lack of SA-dependent activation of the *PR1* gene [18,21]. The *PR1* gene is thought to encode a protein active in defense response and has been used as a marker of SA pathway activation in many studies and in different plant species.

Five independent *TcNPR1* transgenic lines, wild type *Arabidopsis* Col-0 along with the *npr1-2*mutant were sprayed with 1 mM SA, and the expression of *TcNPR1* and *AtPR1* was determined by semi-quantitative RT-PCR 24 hr after induction. Five transgenic lines all showed heterologous*TcNPR1* expression with varied expression levels (Figure 3). As expected, there was no significant up-regulation of the transgene after SA treatment because *TcNPR1* was expressed constitutively from the E12-Ω promoter. The *Arabidopsis PR1* gene showed a very large induction after SA treatment in wild type *Arabidopsis* Col-0 (Figure 3), but there was no up regulation in the *npr1-2*mutant, which is consistent with previous report [69]. There was a small increase in *PR1*expression in the mutant treated with water, which could be expected from plant to plant biological variation. However, the *PR1* gene expression level did not change after SA treatment, as expected for the *npr1-2* mutant. We observed a moderate induction of the *PR1* gene in 3 out of 5 transgenic lines (Line 2, 3 and 4), though the level of induction was not as high as in wild type Col-0. No *PR1*gene induction was observed for transgenic lines 1 and 5. These results suggest that the *TcNPR1*gene can at least partially complement the *Arabidopsis npr1*mutant and act to mediate SA dependent *PR1* gene expression in *Arabidopsis* leaves but it may not act as efficiently as the endogenous NPR1 itself.

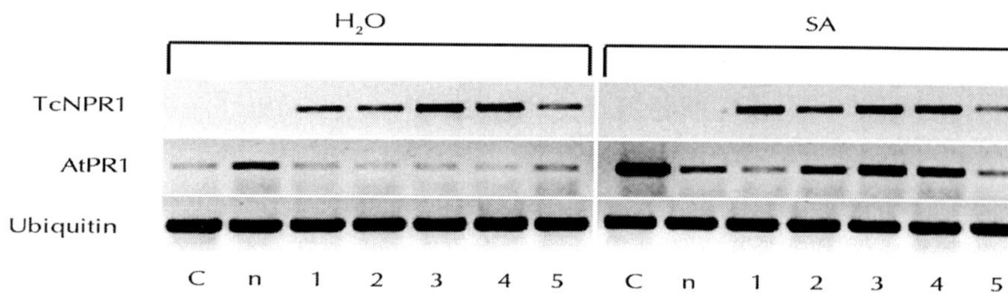

Figure 3: Gene expression of *TcNPR1* and *AtPR1* in transgenic*Arabidopsis npr1-2* lines. Semi-quantitative RT-PCR was performed with cDNA prepared from the leaves of 4-week-old plants of wild type (C),*npr1-2* (n) and 5 independent transgenic *npr1-2* mutant lines overexpressing *TcNPR1* (1-5).*TcNPR1* and *AtPR1* expression were evaluated 24 hrs after 1 mM SA treatment. Water-treated control leaves (left panel) from each genotype were also analyzed. *Arabidopsis Ubiquitin*(*AtUbiquitin*) expression was assayed as a non SA-induced, cDNA loading control.

Another phenotype of the *Arabidopsis npr1* mutation is increased pathogen growth after bacterial infection of leaves [18,21,69]. To test if *TcNPR1* overexpression in *npr1-2* mutant can complement the mutant disease susceptible phenotype, we infected leaves from 5 transgenic lines with*Pseudomonas syringae* pv. tomato DC3000 (*P.s.t.*) by syringe infiltration. The results indicated that the *npr1-2* mutant was more susceptible than Col-0 (Figure 4A) three days after inoculation, exhibiting yellow necrosis similar to previous results [69]. Three transgenic lines overexpressing the *TcNPR1* gene and exhibiting SA dependent *PR1* up-regulation partially restored induced resistance compared to the control *npr1-2* mutant (Figure 4A). Although several yellow necrotic spots were displayed on leaves of the transgenic plants, they did not exhibit severe necrosis or senescence. However, the other two transgenic lines, line 1 and 5, showed necrosis all over the leaves and the tissues were wilted. Water infiltration served as a control to demonstrate that the injection of water alone did not damage the tissues.

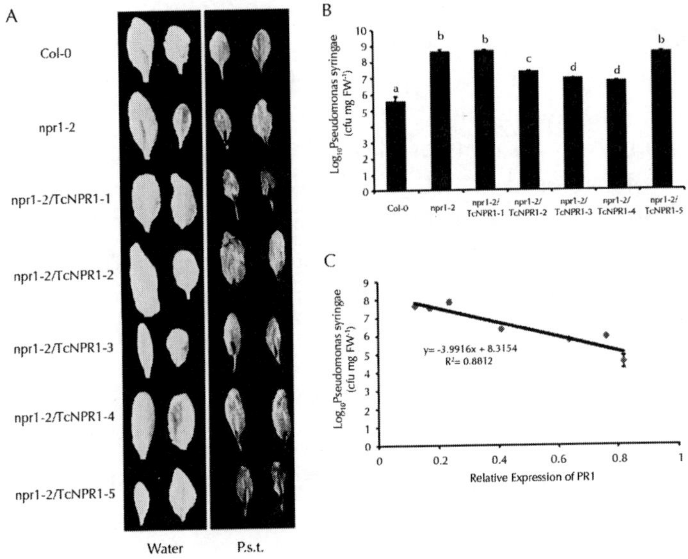

Figure 4: *Pseudomonas syringae* infection assay of transgenic*Arabidopsis npr1-2* mutant lines. A. Disease symptoms on leaves of Col-0, *npr1-2* and five independent lines of *npr1-2* plants transformed with *TcNPR1* (*npr1-2/ TcNPR1*) inoculated with *Pseudomonas syringae* pv. tomato DC3000 (*P.s.t.*) (OD_{600} = 0.002) at three days post inoculation and on leaves of the same seven genotypes infiltrated with water as a control treatment. B. Growth of *P.s.t.* in leaves from Col-0, *npr1-2* and five individual transgenic lines (*npr1-2/ TcNPR1*). Three days after inoculation, leaf disks were collected and bacterial titers were measured. Data represents the means ± standard errors of three biological replicates, each containing three leaf disks from three individual plants. Letters above the histogram indicate statistically significant differences among genotypes ($P < 0.01$) using the single factor ANOVA.C. Correlation of bacterial growth with relative *AtPR1* expression level. Average growth of*Pseudomonas syringae* pv. tomato DC3000 (Figure 4B) and average *AtPR1* gene expression (Figure 3) were evaluated in leaf tissue of Col-0, *npr1-*

2 mutant and five transgenic lines expressing *TcNPR1*. Data was plotted and analyzed by liner regression analysis.

To quantify the disease symptom, bacterial assays were carried out to measure the titer of bacterial on infected leaves. The levels of bacterial in infected *npr1-2* mutant leaf disks increased more than 250 fold as compared to Col-0 controls (Figure 4B). The three transgenic lines overexpressing the *TcNPR1* gene (Line 2, 3 and 4), which exhibited significant up-regulation of the *PR1* after SA treatment, showed a 30 to 100 fold reduction of bacterial growth compared to the*npr1-2* mutant. There was no significant change in bacterial growth rates in leaf disks of the other two transgenic lines tested (Line 1 and 5). To assess the relationship between the level of SA-dependent induction of *PR1* and the degree of bacterial growth in the transgenic lines, we plotted the values as depicted in Figure 4C. A significant negative correlation between SA dependent gene induction and bacterial growth was observed ($R^2 = 0.88$), suggesting that the resistance conferred by *TcNPR1* is via the SA dependent resistance pathway and further supports our hypothesis that TcNPR1 plays a similar function to *Arabidopsis* NPR1 in plant defense response.

Nuclear Translocation of *TcNPR1* after SA Induction

Another hallmark of AtNPR1 function is its nuclear localization in response to treatment with SA[2,25,59,70,71]. To determine if TcNPR1 can also translocate into the nucleus in response to SA in a manner similar to *Arabidopsis* NPR1, we created transgenic *Arabidopsis* plants containing a TcNPR1-EGFP translational fusion and observed the subcellular localization of the fusion protein using confocal microscopy (Figure 5). This construct (35S:*TcNPR1:EGFP*) was stably transformed into the *npr1-2* mutant and we observed the localization of EGFP fusion protein before and 24 hrs after SA treatment in both leaf and root cells of four independent transgenic lines. We observed no EGFP fluorescence in negative control plants transformed with the identical vector lacking the TcNPR1-EGFP fusion gene (Figure 5A and 5B). As an additional control, transgenic plants overexpressing EGFP without a fusion to TcNPR1 were imaged, and we observed strong fluorescence in both cytoplasm and nucleus with no localization changes after SA treatment. A final control consisted of a construct designed for the overexpression of the *Arabidopsis* NPR1 protein translationally fused to EGFP (35S:*AtNPR1*:*EGFP*). Consistent with the findings of others [25,59], we observed an extremely strong nuclear translocation of the fusion protein in leaf guard cells and in root cells 24 hrs after SA treatment.

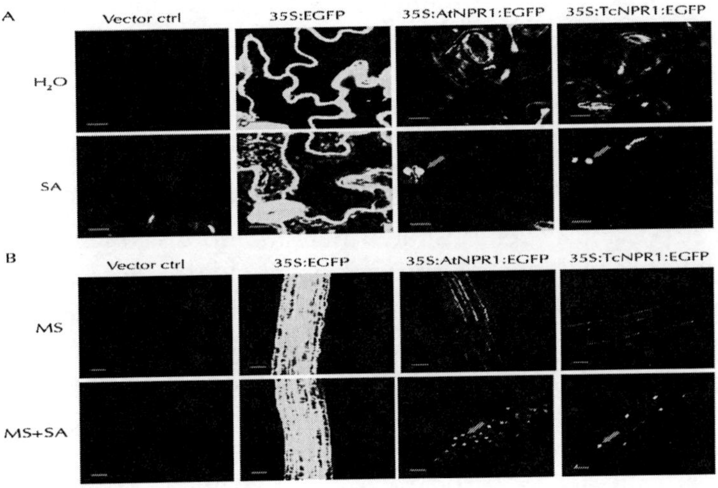

Figure 5: Nuclear localization of TcNPR1-EGFP in transgenic*Arabidopsis* plants in response to SA. A. Confocal images of EGFP fluorescence in *Arabidopsis* leaves of 4-week-old soil-grown plants 24 hrs after H_2O (upper images) or 1 mM SA (lower images) treatment. All images were taken at the same magnification and exposure times. Arrows indicate the accumulation of green fluorescence in guard cell nuclei after SA treatment. Scale bar, 10 μm. B. Confocal images of EGFP fluorescence in*Arabidopsis* roots from 10-day-old seedlings grown on MS (upper images) or MS supplemented with 0.5 mM SA (lower images). All images were captured using the same exposure settings. Arrows indicate the accumulation of EGFP in nuclei of root cells after SA treatment. Scale bar, 30 μm. Samples from transgenic plants generated with pCAMBIA1300 (vector ctrl) was used as negative control and samples from transgenic plants expressing 35S:EGFP served as positive control in A and B.

The TcNPR1-EGFP fusion protein appeared to be evenly distributed in cytoplasm of leaf guard cells from water-treated 4-week-old soil grown plants, however, the protein accumulated moderately in guard cell nucleus 24 hours after SA application (Figure 5A, red arrow). Similarly, a modest level of nuclear translocation could also be observed in the root cells from 10-day-old seedlings grown on MS medium supplemented with 0.5 mM SA (Figure 5B). Although protein translocation of TcNPR1 is of lesser extent than observed with the *Arabidopsis* NPR1-EGFP protein based on reduced nuclear fluorescence observed in TcNPR1-EGFP transgenic plants, our results taken together indicate that TcNPR1, like *Arabidopsis* NPR1, can translocate into nucleus after SA induction and participate in the induction of defense related gene expression.

TcNPR1 and SA-JA Crosstalk

It has been previously demonstrated that *Arabidopsis* NPR1 can mediate the antagonism between SA and jasmonic acid (JA) by suppressing JA-responsive genes [27,34,35], suggesting that it plays an important role in fine tuning the cross-talk between different regulatory pathways. To explore the role of TcNPR1 in cross-talk, we tested the effect of SA and JA treatments on defense gene expression in wild type Col-0, *npr1-2* mutant and five independent 35S:TcNPR1 transgenic *Arabidopsis* lines. Semi-quantitative RT-PCR showed that all five lines carrying the cacao transgene expressed *TcNPR1* at moderate levels, and these did not change much during hormone treatments (Figure 6A). Exogenous application of 1 mM SA activated *PR1* in Col-0 and three transgenic lines, but not in *npr1-2* mutant. Additionally, 48 hrs after treatment with 0.1 mM methyl jasmonate (MeJA) in 0.015% Silwet L-77, two well established MeJA inducible genes (*VSP2* and *PDF1.2*) were up-regulated in wild-type plant and in *npr1-2* mutant, consistent with previous reports [34,72]. Two DNA bands were detected in some of the *PDF1.2* PCR products, and we determined that the smaller molecular weight band resulted from cDNA amplification and the large fragment resulted from amplification of genomic DNA (data not shown). As predicted, all five 35S:TcNPR1 transgenic lines exhibited levels of increased *VSP2* and *PDF1.2* that were similar to those seen in Col-0 plants. Upon treatment with a combination of 1 mM SA and 0.1 mM MeJA in 0.015% Silwet L-77, *PR1* was expressed at a level similar to seen when plants were treated with SA alone, indicating that MeJA had no effects on SA-responsive *PR1* expression. Both *VSP2* and *PDF1.2* expressed at significantly lower levels in Col-0 compared to that in *npr1-2* mutant after SA and MeJA combined treatment, demonstrating the function of AtNPR1 in antagonistic repression of JA-responsive genes. All five transgenic lines containing *TcNPR1* gene displayed reduced expression levels of JA-responsive gene expression upon SA and JA combined treatment compared to *npr1-2* mutant, suggesting that TcNPR1 can also mediate SA-JA cross-talk in a manner similar to AtNPR1.

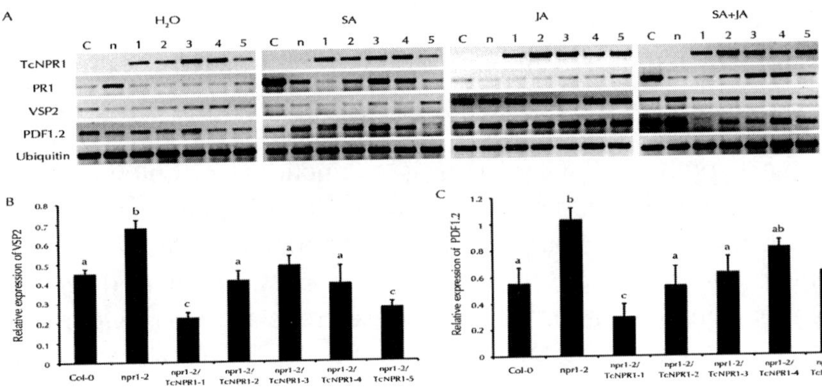

Figure 6: Gene expression of SA- and JA-responsive genes in transgenic *Arabidopsis npr1-2* mutants. A. Semi-quantitative RT-PCR was performed with cDNA prepared from leaves of 4-week-old plants of wild type(C), *npr1-2* (n) and 5 independent transgenic *npr1-2* mutant lines overexpressing *TcNPR1* (1-5). The expression of *TcNPR1, AtPR1, AtVSP2* and *AtPDF1.2*was evaluated 48 hrs after treatment with water control, 1 mM SA water solution alone, 0.1 mM MeJA alone in 0.015% Silwet L-77 and the combination of 1 mM SA and 0.1 mM MeJA in 0.015% Silwet L-77. *AtUbiquitin* was used as a cDNA loading and normalization control. B. The intensity of *AtVSP2* and *AtPDF1.2* RT-PCR gel bands in Figure 6A were quantified by ImageQuant software for total pixel intensity and the expression levels were normalized by*AtUbiquitin*. The bar charts represent the means ± standard errors of relative expression value of *AtVSP2* and *AtPDF1.2* following 48 hrs treatment of SA-MeJA combination of three biological replicates. Letters above the bar chart indicate statistically significant differences among genotypes ($P < 0.05$) determined by single factor ANOVA.

To quantify the expression of *VSP2* and *PDF1.2* after the treatment of the combination of SA and MeJA, we measured the band intensity as above (Figure 6B). The data was normalized to an*Ubiquitin* control for loading effects. The relative expression levels of *VSP2* and *PDF1.2* were significantly decreased in *TcNPR1* expressing transgenic lines compared to *npr1-2* mutant ($P < 0.05$), a pattern similar to wild-type Col-0, suggesting that *TcNPR1* restored the *npr1* phenotype. These data support our hypothesis that TcNPR1 may play a role in mediating SA-JA cross talk as does *Arabidopsis* NPR1.

DISCUSSION

We have isolated an *NPR1* homologous gene from the tropical tree, *Theobroma cacao*, and have generated transgenic *Arabidopsis npr1-2* mutant lines overexpressing *TcNPR1*. All of our results are consistent with the hypothesis that *TcNPR1* is a functional orthologue of the well characterized *Arabidopsis* gene. *TcNPR1* complemented each of the major *Arabidopsis npr1-2* mutant phenotypes that were tested. Over-expression of *TcNPR1* in the *npr1-2* mutant conferred *PR1* up-regulation after SA treatment and increased resistance to *Pseudomonas syringae* pv. tomato DC3000 (Figure 3 and 4A, B). *TcNPR1* was shown to be translocated into the nucleus in response to SA and to participate in SA-JA cross talk regulation (Figure 5 and 6). In our data, we found that transgenic lines 1 Line exhibited reduced complementation in SA-induced *PR1* expression and disease resistance (Figure 3 and 4), while at the same time same two lines efficiently mediated crosstalk between SA and JA (Figure 6). In previous studies, the activation of defense related genes was shown to involve the nuclear translocation of NPR1 [59] while the crosstalk between SA and JA signaling was shown to be mediated by cytosolic NPR1 [34], thus it appears that very different mechanisms exist for these two functions of NPR1. The differential efficiencies of complementation of TcNPR1 we observed may reflect these different mechanisms. It is well known that positional effects (differential transgene transcription levels due to different genomic insertion sites in individual transgenic events) can have a large effect on protein expression levels. As suggested by RNA expression levels of the different TcNPR1 expressing transgenic lines (Figure 3), lines 1 and 5 may have lower protein expression than lines 2-4. It seems plausible that the differential complementation of the two NPR1 functions resulted from the differences in expression levels, potentially as a result of different protein accumulation levels in the cytosol vs nuclear compartments. Consistent with this idea, only the higher levels of expression seen in lines 2-4 was sufficient to complement the nuclear gene induction function, but the levels of expression were high enough in all lines to complement the cytosolic SA/JA crosstalk function.

In all, our results demonstrate a high degree of evolutionary and functional conservation of NPR1 from the Brassicales to the Malvales. NPR1 is also conserved in species as diverse as grapevine[14], tomato [12], apple [37], banana [73], cotton [74], tobacco [8] and rice [11]. This high degree of functional conservation suggests that NPR1 function evolved very early in the development of higher plants and that it plays a very critical role in plant development and reproductive success.

Little is known about the mechanisms of defense signaling in cacao. Our data suggests that the central mechanisms operative in *Arabidopsis* are likely to be conserved in cacao. At a minimum, our data suggests that the mechanisms and molecules that interact with NPR1 during SA and JA signaling and nuclear translocation are also conserved in cacao. If this were not the case, we would not expect the cacao NPR1 protein to function normally in *Arabidopsis*. However, the cacao protein in some cases only partially restored function of the *npr1* mutant, which is likely the result of transgene expression level differences compared to the endogenous gene and/or partial molecular incompatibility with its interacting protein partners. It is possible that the binding affinities between the cacao NPR1-interacting proteins are reduced as compared to the endogenous *Arabidopsis* coevolved partners. Partial complementation has commonly been observed in heterologous complementation analysis in many other systems [75-77].

Further investigation is needed to explore the entire defense response pathway in *Theobroma cacao* and to understand the similarities and differences with *Arabidopsis* overall. For example, our expression data shows that *TcNPR1* can be up-regulated only at 4 mM SA treatment but not at lower concentrations, which is higher than the optimal level of 1 mM in *Arabidopsis* as previously indicated [69]. It would be interesting to test the endogenous SA level of cacao and to determine dose response dynamics in various tissues and during different stages of development. Another area of interest is to identify and characterize the downstream targets of *TcNPR1* and to compare them to the approximately 2,248 genes that are regulated by NPR1 during systemic acquired resistance in *Arabidopsis* [78]. Surveying these genes in cacao could reveal interesting differences in the defense responses unique to this tropical tree relative to *Arabidopsis*. Furthermore,*Arabidopsis* NPR1 has been shown to interact with several different proteins such as the TGA transcription factors [16,28,33,70,79]. Thus another area of interest is to isolate *TcNPR1* interacting cacao proteins, which will further enhance our knowledge of this pathway in cacao. We are also interested in studying other NPR1-like genes of cacao and the recent completion of a draft cacao genome sequence has led to the identification of three additional NPR1-like cacao genes[80].

Plant diseases, especially pathogenic fungi, are estimated to cause about 30-40% yield loss on cacao annually [41,81], and thus disease resistance is of substantial interest to cacao breeders. Our findings can be utilized in several approaches to help develop varieties of cacao with enhanced disease resistance. The sequence of the *TcNPR1* gene could possibly be used to develop molecular markers and probes that can be employed to select disease resistant varieties with specific allelic variations. Interestingly, the major quantitative trait locus (QTLs) for witches' broom disease resistance is tightly linked to the *TcNPR1* gene [80], thus the *TcNPR1* gene serves as a key candidate gene for generation of molecular markers that can be used for marker assisted selection of new disease resistant varieties. In addition, *TcNPR1* expression levels could be modified in transgenic cacao varieties to develop broad-spectrum disease resistance. This approach has already been successful in several species but to our knowledge, has not yet been deployed in commercial production for any species. However, consumer and industry reluctance to accept transgenic plant technology remains a formidable barrier to development of any transgenic cacao varieties for commercialization.

CONCLUSIONS

The isolation of the *TcNPR1* gene and its heterologous complementation in *Arabidopsis* allowed us to rapidly characterize the function of this defense-related gene. The up-regulation of *PR1* and increased bacterial resistance in transgenic *Arabidopsis npr1-2* mutants strongly supported that*TcNPR1* is a functional ortholog of Arabidopsis *NPR1*, and vital component in SA-dependent signaling pathway in *Theobroma cacao*. Our results provide potential opportunities to enhance disease resistance in this crop species through conventional breeding or biotechnological approaches. Further investigation is needed to identify the TcNPR1 interacting transcription factors and their downstream targets in cacao and to reveal further details of the molecular mechanisms of the role TcNPR1 plays as a central mediator of the plant defense response.

METHODS

Full-Length cDNA cloning by Degenerate PCR

NPR1 cDNA sequences from *Arabidopsis* (U76707), *Brassica napus* (AF527176), and *Carica papaya*(AY550242) were aligned using the ClustalW program v1.8 [82]. Degenerate primers (TcNPR1dg-5′, TATTGTCAARTCTRATGTAGAT; TcNPR1dg-3′, GAARAAYCGTTTCCCKAGTTCCAC) were designed to regions highly conserved among all three sequences.

Total RNA was isolated from cacao leaves from variety Scavina6 as previously described [53]. Cacao leaf cDNA was synthesized using the SMART RACE cDNA Amplification Kit (Clontech Laboratories Inc., Mountain view, CA http://www.clontech.com/) according to the manufacturer's instructions.

PCR reactions were performed using cacao leaf 2.5 µl cDNA from first strand synthesis from SMART RACE cDNA Amplification Kit, 10 µl Redi-prime PCR mix (GeneChoice, Inc., Frederick, MD) and 5 µM of the above degenerate primers. Following denaturation (94° for 5 min.), PCR was performed for 32 cycles using the following condition (94° for 30 sec., 45° for 30 sec, 72° for 1 min.), followed by a 5 min. extension at 72°. PCR products were resolved on 1% agarose gels, purified with the GENECLEAN II Kit (Q-Biogene Inc., Solon OH) and cloned into the pGEM-T-Easy vector (Promega Corporation, Madison WI) according to the manufacturer›s instructions. DNA sequencing was performed at the Penn State Genomics Core Facility using an ABI Hitachi 3730XL DNA Analyzer. The resulting clone was designated as pGEM-TcNPR1.

Genomic DNA Cloning by BAC Library Screening

Theobroma cacao BAC filter arrays constructed using genomic DNA from genotype LCT-EEN 37 were purchased from the Clemson University Genomic Institute http://www.genome.clemson.edu/. Filter arrays were blocked for 4 hours at 60°C in a solution containing 1% BSA, 1 mM EDTA, 7% SDS, and 0.25 M sodium phosphate. PCR generated *TcNPR1* cDNA fragment labeled with ^{32}P dCTP using the MegaPrimer DNA Labeling System (GE Healthcare, Buckinghamshire, UK) according to the manufacturer's instructions was added and hybridized overnight at 60°C. The next day, the filter arrays were washed twice in 2× sodium chloride/sodium citrate (SSC), 0.5% sodium dodecyl sulfate (SDS) for 20 minutes at 60°C. Radiographic imaging was performed via storage phosphor imaging (Molecular Dynamics, http://www.mdyn.com/). After filter alignment and clone number identification, a BAC clone (2K13) containing a putative *TcNPR1* fragment was obtained from a frozen stock. The sequence of *TcNPR1* genomic fragment was acquired by series sequencing of the BAC clone from ATG start codon. Sequencing primer was designed based on the*TcNPR1* cDNA at the first round and following series primers were designed based on the known sequence resulting from previous sequencing. Introns were identified by aligning the genomic sequence and full length cDNA using SPIDEY software http://www.ncbi.nlm.nih.gov/spidey/ The same strategy was applied to clone the 1.1 kb promoter region upstream of the ATG. Forward and reverse sequencing was also performed to validate the sequence.

For sequence verification the *Arabidopsis* NPR1 protein sequence (At1g64280) and putative cacao NPR1 protein sequences (genbank accession HM117159) were aligned using the ClustalW program v1.83 [82]. The TcNPR1 protein sequence was analyzed for potential functional sites by querying the Simple Modular Architecture Research Tool (SMART) database http://smart.embl-heidelberg.de/

Semi-Quantitative RT-PCR Analysis of *TcNPR1* Expression in Cacao Tissues

Total RNA was isolated from Scavina6 leaves stages A, C and E (corresponding to stages YR, IG, MG respectively, as described in

[83]), open flowers, unopened flowers, roots, exocarps and seeds as previously described [53]. For each tissue, three biological replicates were collected and analyzed. Cacao cDNA was synthesized in a final volume of 25 μl from 2 μg of total cacao RNA using M-MLV reverse transcriptase (New England Biolabs, Inc., Ipswich, MA). RNA and 0.5 μg oligo(dT) were added to sterile water to final volume of 18 μl. The mixture was then incubated at 70°C for 5 min, chilled on ice, which was followed by adding 10× reverse transcription buffer (New England Biolabs, Inc., Ipswich, MA), 0.1 M fresh made DTT and 10 mM dNTP. The mixture was further incubated at 42°C for 2 min, followed by incubation at 42°C for 1 hr with 10 units of reverse transcriptase MMLV (New England Biolabs, Inc., Ipswich, MA). The reaction was terminated at 70°C for 15 min.

Semi-quantitative RT-PCR was performed using intron-spanning primers for *TcNPR1* (TcNPR1RT-5′: ATGGATTCCCGTCTGGAACTTGGT; TcNPR1RT-3′: TCTGGAGTGTCATTTCCTCCGCAT) and *TcActin*(CL33contig2 in Esttik Databasehttp://esttik.cirad.fr/usedasaninternalnormalizationandcDNA loading control (TcActinRT-5′: AGCTGAGAGATTCCGTTGTCCAGA and TcActinRT-3′: CCCACATCAACCAGACTTTGAGTTC). RT-PCR reactions were set up using 1 μl of 1/2 diluted cDNA and 5 μM of the *TcNPR1* or *TcActin* primers. Titration of cycles was carried out and it was determined that the PCR amplification of *TcNPR1* was within its linear range at 27 cycles using the following condition: 94°C for 30 sec., 56°C for 30 sec, 72°C for 1 min. Similarly, PCR of *TcActin*was performed under non-saturation conditions within the linear range (22 cycles at 94°C for 30 sec., 55°C for 30 sec, 72°C for 1 min). *TcActin* served as a cDNA loading control.

SA Treatment of Cacao Seedlings

The leaves of two to three-month old cacao plants generated by rooted cuttings from two different genotypes (ICS1 and Scavina6) were sprayed with salicylic acid (SA) dissolved in water at three different concentrations, 1 mM, 2 mM and 4 mM. Control plants were treated with water. Plants were grown in a greenhouse under conditions previously described [53] and leaf tissue from fully expanded young leaves (developmental stage C, corresponding to stage IG in [46]) was harvested at 24 hrs after treatment and frozen in liquid nitrogen. Total

RNA was isolated and cDNA was synthesized as described above. For each genotype and each treatment, three biological replicates were collected. Semi-quantitative RT-PCR and expression analysis were performed to assay the levels of *TcNPR1* expression as described above. The PCR products were analyzed on 1% agarose gel, stained with ethidium bromide. The expression values of *TcNPR1* and *TcActin* were quantified using ImageQuant software (Molecular Dynamics, Amersham Bioscience) as described in [84] and relative expression of *TcNPR1* was calculated by comparing with the expression of *TcActin*.

Transgenic *Arabidopsis* Mutant Complementation Assay

All binary plant transformation vectors were constructed by incorporating the genes of interest into pCAMBIA-1300 binary transformation vector containing plant selectable marker for hygromycin resistance [85].

Binary Vector p35S:TcNPR1 - The *TcNPR1* coding sequence fragment was generated by PCR using pGEM-TcNPR1 as described above and included *Xma*I and *Not*I restriction sites at the 5′- and 3′-ends respectively (TcNPR1-5′-*Xma*I, CCCGGGATGGATAACAGAAATGGCTT; TcNPR1-3′-*Not*I, GCGGCCGCTTGCATTAGGCCTATGGTCTA). The fragment was cloned into pGEM T-Easy (Promega Corporation, Madison WI) according to the manufacturer's instructions and sequenced for integrity. The *TcNPR1* coding sequence was then cloned into the *Xma*I and *Not*I sites of an intermediate cloning vector (pE2113) between E12-Ω promoter [86] and 35SCaMV terminator. A 3 kb restriction fragment containing *TcNPR1* gene cassette was excised from pE2113 using *Pvu*II and ligated into the *Sma*I site of pCAMBIA-1300.

Ligations were performed overnight at 16° with 3Units of T4 DNA ligase (Promega Corporation, Madison WI).

Binary vector p35S:AtNPR1 - The *AtNPR1* coding sequence fragment was generated by PCR using the *AtNPR1* cDNA clone U13446 from Arabidopsis Biological Resource Centerhttp://www.biosci.ohio-state.edu/~plantbio/Facilities/abrc/abrchome.htm and included*Nco*I sites at the 5′- and 3′-ends (AtNPR1-5′-*Nco*I, CCATGGACACCACCATTGATGGATTC; AtNPR1-3′-*Nco*I, CCATGGTCCGACGACGATGAGAGAGTTTACG). The PCR fragment was cloned into pGEMT-Easy and sequenced. The resulting intermediate plasmid was designated pGEM-AtNPR1. The*AtNPR1* coding sequence was then excised by *Nco*I from pGEM-AtNPR1, and blunt-end cloned into pE2113 between E12-Ω promoter [86] and 35SCaMV terminator as *Xma*I and *Not*I fragment. Contently 3.1 kb fragment containing the *AtNPR1* gene cassette was obtained by digestion with *Pvu*II, and blunt-end ligated into the *Sma*I site of pCAMBIA-1300.

Binary Vector p35S:TcNPR1:EGFP - The cassette of E12-Ω promoter and EGFP on the intermediate cloning vector pE2113 was cloned into *EcoR*I and *Hind*III sites of pCambia1300. The resulting binary vector was designated pXCGH. PCR generated *TcNPR1*fragment, including *Sma*I and *KpnI*sites at the 5′- and 3′-ends (TcNPR1-5′-*Sma*I, CCCGGGATGGATAACAGAAATGGCTT; TcNPR1_3′-*Kpn*I, GGTACCGACCGCCCCTACCACTACCAGTTAG) was first cloned into pGEM T-Easy (pGEM-TcNPR1-EGFP). The sequence was verified, the DNA fragment was excised with *SmaI* and *KpnI* and blunt ends ligated into the blunt-ended *NcoI* site of pXCGH positioned between the E12-Ω promoter and at the 5›end of the EGFP coding sequence to generate the binary vector p35S:TcNPR1:EGFP.

Binary vector p35S:AtNPR1:EGFP - The pGEM-AtNPR1 containing *AtNPR1* coding sequence was digested with *Nco*I and the fragment was ligated into the *Nco*I site of pE2113 as described above. The 3.6 kb fragment containing the *AtNPR1*- EGFP fusion gene cassette was digested with *Sal*I and*Eco*RI and cloned into the *Sal*I and *Eco*RI sites of pCAMBIA-1300.

Arabidopsis **Transformation**

The binary vectors described above were introduced into *Agrobacterium tumefaciens* strain AGL1 by electroporation, as previously described in [87]. *Arabidopsis* Col-0 plants were grown in a Conviron growth chamber (Model No. MTPS144) maintained at 22°C, under a 16:8::L:D cycle. Light intensity was maintained at 200 μM/m^2·s with Octron 4100K Ecologic bulbs (Sylvania, Danvers MA). To increase the number of inflorescences, plants were cut back after bolting, and allowed to re-grow. The floral dip method was used to transform *Arabidopsis* as described previously [88]. Briefly, Agrobacterium cultures were grown at 25° on a platform shaker (200 rpm) to an OD_{600} = 1.2. Cells were centrifuged at 1,500 × *g* for 6 minutes and re-suspended in 300 mls of a solution containing 2.15 g L^{-1} MS salts, 5% sucrose, 0.02% Silwet-77. The flowers were dipped in the solution for three seconds, domed to remain humidity and covered with black cloth. The cloth was removed the next day and plants were regularly watered until seed maturation.

Following seed set, seeds were collected from nine plants for each independent transgenic event. Seeds from 5 individual lines were soaked in 0.1% Tween-20 for 2 minutes and sterilized with 50% bleach for 10 minutes at room temperature. Seeds were then washed five times with 1 ml of sterile water. To select for transformants, seeds were planted on 1/2 MS media, agar plates (pH 5.7) supplemented with 25 µg ml^{-1} hygromycin B. Plates were place in a Conviron growth chamber under the same light and temperature conditions as above. After 10 days, germinated seedlings were examined for leaf development and root elongation. Those seedlings that showed root elongation were transferred to soil and allowed to grow. Transformations were performed with the following vectors: p35S:TcNPR1, p35S:AtNPR1, p35S:TcNPR1:EGFP, and p35S:AtNPR1:EGFP constructed as described above, and control vectors p35S: EGFP (pGH00.0126, [89])and pCambia 1300.

Salicylic Acid (SA) *Arabidopsis* Induction Assay

Four week-old wild type *Arabidopsi*s Col-0 and *npr1-2* mutants and five independent transgenic lines growing in soil were sprayed with 1 mM SA, along with water-treated control plants. Three biological replicates, each containing leaves from 5 individual plants were collected 24 hrs after treatment. Total RNA was isolated from treated and control samples using RNeasy plant mini kit (QIAGEN, Valencia CA). cDNA was generated as described above. Semi-quantitative RT-PCR was performed as described above to measure the expression of *TcNPR1* and *AtPR1*. *Arabidopsis Ubiquitin* served as cDNA loading and normalization control. The following primer sets and conditions were employed:

TcNPR1-5′: ATGGATTCCCG TC TG GA ACTTGGT; TcNPR1-3′: TCTG GAGT GTC ATTTCCT CCGCAT (27 cycles of 94°C for 30 sec., 56°C for 30 sec., 72°C for 1 min). AtPR1-5′: CT CGA AAG CTCAAGATAGCCCACA; AtPR1-3′: CTTCTC GTTCACATAATTCCCACG

(25 cycles of 94°C for 30 sec., 54°C for 30 sec., 72°C for 1 min). Ubiquitin-5′: ACCGGCAAGACCATCACTCT; Ubiquitin-3′: AGGCCTCAACTGGTTGCTGT (22 cycles of 94°C for 30 sec., 54°C for 30 sec., 72°C for 1 min). The conditions of PCR were determined by cycle titration to avoid saturating conditions. The relative expression levels were determined as described above.

Pseudomonas Syringae Infection Assay of *Arabidopsis* Transgenic Plants Overexpressing *NPR1* Genes

Pseudomonas syringae pv. tomato DC3000 (*P.s.t.*) was grown on Difco Pseudomonas agar (PA) (Becton, Dickinson and Company, http://www.bdbioscience.com/) supplemented with rifampicin (100 μg ml-1) and kanamycin (25 μg ml-1) at 25°C for 48 hrs. Cells were scraped from plates using a bacterial inoculating loop and re-suspended in water. Plant infection assays and bacterial growth assays were carried out as described previously in [90]. Five individual transgenic*npr1-2* mutant overexpressing *TcNPR1* coding sequence were infected with *P.s.t.* at OD_{600} = 0.002. Briefly, three days after inoculation leaf disks from treated leaves of 2 independent replicate plants were pooled for a single sample. Data represents means ± SE (cfu/mg FW) of three biological replicates per treatment and statistical differences were determined by Single factor ANOVA analysis.

Nuclear Translocation of TcNPR1 in Transgenic *Arabidopsis* Plants

For observations of green leaves, four week-old soil-grown transgenic plants containing one of transgenes 35S:AtNPR1:EGFP, 35S:TcNPR1:EGFP, 35S:EGFP and plants transformed with empty binary vector pCambia 1300 were sprayed with either a 1 mM solution of SA in water or water. For root observations, control and transgenic seed were germinated on MS agar or MS agar supplemented with 0.5 mM SA [59] and seedlings were grown for 10 days. Leaves and roots were placed in a drop of water on a standard microscope glass slide and overlaid with a cover slip. The samples were imaged with an inverted

Olympus FV1000 Laser Scanning Confocal Microscope (Olympus America Inc., Melville, NY). For imaging EGFP, tissues were excited with a blue argon laser (488 nm) and emission wavelengths of 500-600 nm were detected through a variable bandpass filter positioned in front of the photomultiplier tube. Tissues were observed using 40× and 10× objectives for leaf cells and root cells, respectively, each with numerical apertures and 1.15. FV10-ASW version 1.6 software (OLYMPUS, Pittsburgh, PA) was used to collect images, select slices, and create intensity projections over the Z axis.

SA and JA Combination Treatment of *Arabidopsis* Transgenic Plants Overexpressing *TcNPR*1

Four weeks old soil-grown wild type *Arabidopsi*s Col-0, *npr1-2* mutants and five independent transgenic lines containing p35S:TcNPR1 were sprayed with a combination of 1 mM SA and 0.1 mM MeJA in 0.015% Silwet L-77. Plants treated with 1 mM SA alone in water, 0.1 mM MeJA alone in 0.015% Silwet L-77 and water with 0.015% Silwet L-77 served as control treatment. Three biological replicates each consisting of leaves from 5 individual plants were collected at 48 hrs after treatment, total RNA was isolated, cDNA was synthesized and semi-quantitative RT-PCR was performed as described above to determine the transcripts level of *TcNPR1* and *AtPR1*. For expression analysis of *VSP2* and *PDF1.2*, following primer sets and conditions were used to maintain the reaction in its linear amplification range. *VSP2* Forward: TACGGTCTCGGCATCCGTTC;*VSP2* Reverse: CCTCAAGTTCGAACCATTAGGCT (21 cycles of 94°C for 30 sec., 58°C for 30 sec., 72°C for 1 min). *PDF1.2* Forward: TCATCATGGCTAAGTTTGCTTCCATC; *PDF1.2* Reverse: TGTCATAAAGTTACTCATAGAGTGAC (27 cycles of 94°C for 30 sec., 60°C for 30 sec., 72°C for 1 min). The PCR products were analyzed on 1% agarose gel, stained with ethidium bromide. The expression values of *AtVSP2* and *AtPDF1.2* were quantified using ImageQuant software (Molecular Dynamics, Amersham Bioscience) as described in [84] and relative expression of two genes was calculated by comparing with the expression of *AtUbiquitin*.

AUTHORS' CONTRIBUTIONS

ZS performed most of the experiments, *ie*, sequence analysis, gene expression studies, phenotypic analysis of transgenic *Arabidopsis* plants, confocal microscopy observations and drafted the manuscript. SNM participated in the design of the study, directed the transformation vector construction and transgenic lines generation, and participated in drafting of the manuscript. YL participated in transgenic *Arabidopsis* plants analysis and helped to analyze the sequence. JV cloned the *TcNPR1* gene. MJG conceived the study, drafted the manuscript and gave advice on experimental design, data analysis and execution. All authors read and approved the final manuscript.

ACKNOWLEDGEMENTS

We would like to thank Nicole Zembower at cytometry facility for providing help for confocal microscopy imaging. Thanks to the people in the Guiltinan lab, especially Ann Young and Sharon Pishak for their technical assistance in construction the transformation vectors and generation of transgenic plants. This work is supported in part by The Pennsylvania State University, The Huck Institutes of Life Sciences and American Research Institute Penn State Endowed Program in the Molecular Biology of Cacao.

REFERENCES

1. Heath MC: Hypersensitive response-related death. *Plant Mol Biol* 2000, 44(3):321-334.
2. Durrant WE, Dong X: Systemic acquired resistance. *Annu Rev Phytopathol* 2004, 42:185-209.
3. Attaran E, Zeier TE, Griebel T, Zeier J: Methyl salicylate production and jasmonate signaling are not essential for systemic acquired resistance in Arabidopsis. *Plant Cell* 2009, 21(3):954-971
4. Mur LA, Kenton P, Lloyd AJ, Ougham H, Prats E: The hypersensitive response; the centenary is upon us but how much do we know? *J Exp Bot* 2008, 59(3):501-520.

5. Dong X: NPR1, all things considered. *Curr Opin Plant Biol* 2004, 7(5):547-552.

6. Mukhtar MS, Nishimura MT, Dangl J: NPR1 in plant defense: It's not over 'til It's turned over. *Cell* 2009, 137(5):804-806.

7. Cao H, Bowling SA, Gordon AS, Dong X: Characterization of an Arabidopsis mutant that is nonresponsive to inducers of systemic acquired resistance. *Plant Cell* 1994, 6(11):1583-1592

8. Liu Y, Schiff M, Marathe R, Dinesh-Kumar SP: Tobacco *Rar1, EDS1* and *NPR1/NIM1* like genes are required for N-mediated resistance to tobacco mosaic virus. *Plant J* 2002, 30(4):415-429.

9. Fitzgerald HA, Chern MS, Navarre R, Ronald PC: Overexpression of *(At)NPR1* in rice leads to a BTH- and environment-induced lesion-mimic/cell death phenotype. *Mol Plant Microbe Interact* 2004, 17(2):140-151.

10. Vlot AC, Klessig DF, Park SW: Systemic acquired resistance: the elusive signal(s). *Curr Opin Plant Biol* 2008, 11(4):436-442.

11. Yuan Y, Zhong S, Li Q, Zhu Z, Lou Y, Wang L, Wang J, Wang M, Yang D, He Z: Functional analysis of rice *NPR1-like* genes reveals that *OsNPR1/NH1* is the rice orthologue conferring disease resistance with enhanced herbivore susceptibility. *Plant Biotechnol J* 2007, 5(2):313-324.

12. Chen YY, Lin YM, Chao TC, Wang JF, Liu AC, Ho FI, Cheng CP: Virus-induced gene silencing reveals the involvement of ethylene-, salicylic acid- and mitogen-activated protein kinase-related defense pathways in the resistance of tomato to bacterial wilt. *Physiol Plant* 2009, 136(3):324-335.

13. Makandar R, Essig JS, Schapaugh MA, Trick HN, Shah J: Genetically engineered resistance to *Fusarium* head blight in wheat by expression of *Arabidopsis NPR1*. *Mol Plant Microbe Interact* 2006, 19(2):123-129.

14. Le Henanff G, Heitz T, Mestre P, Mutterer J, Walter B, Chong J: Characterization of *Vitis vinifera NPR1* homologs involved in the regulation of pathogenesis-related gene expression. *BMC Plant Biol* 2009, 9:54.

15. Gaffney T, Friedrich L, Vernooij B, Negrotto D, Nye G, Uknes S, Ward E, Kessmann H, Ryals J: Requirement of salicylic acid for the induction of systemic acquired resistance. *Science* 1993, 261(5122):754-756.

16. Rochon A, Boyle P, Wignes T, Fobert PR, Despres C: The coactivator function of Arabidopsis NPR1 requires the core of its BTB/POZ domain and the oxidation of C-terminal cysteines. *Plant Cell* 2006, 18(12):3670-3685
17. Loake G, Grant M: Salicylic acid in plant defence--the players and protagonists. *Curr Opin Plant Biol* 2007, 10(5):466-472.
18. Shah J, Tsui F, Klessig DF: Characterization of a salicylic acid-insensitive mutant (*sai1*) of *Arabidopsis thaliana*, identified in a selective screen utilizing the SA-inducible expression of the *tms2* gene. *Mol Plant Microbe Interact* 1997, 10(1):69-78.
19. Delaney TP, Friedrich L, Ryals JA: *Arabidopsis* signal transduction mutant defective in chemically and biologically induced disease resistance. *Proc Natl Acad Sci USA* 1995, 92(14):6602-6606.
20. Glazebrook J, Rogers EE, Ausubel FM: Isolation of *Arabidopsis* mutants with enhanced disease susceptibility by direct screening. *Genetics* 1996, 143(2):973-982.
21. Ryals J, Weymann K, Lawton K, Friedrich L, Ellis D, Steiner HY, Johnson J, Delaney TP, Jesse T, Vos P, *et al.*: The Arabidopsis NIM1 protein shows homology to the mammalian transcription factor inhibitor I kappa B. *Plant Cell* 1997, 9(3):425-439
22. Cao H, Li X, Dong X: Generation of broad-spectrum disease resistance by overexpression of an essential regulatory gene in systemic acquired resistance. *Proc Natl Acad Sci USA* 1998, 95(11):6531-6536
23. Stogios PJ, Downs GS, Jauhal JJ, Nandra SK, Prive GG: Sequence and structural analysis of BTB domain proteins. *Genome Biol* 2005, 6(10):R82.
24. Baldwin AS: The NF-kappa B and I kappa B proteins: new discoveries and insights. *Annu Rev Immunol* 1996, 14:649-683.
25. Mou Z, Fan W, Dong X: Inducers of plant systemic acquired resistance regulate NPR1 function through redox changes. *Cell* 2003, 113(7):935-944.
26. Pieterse CM, Van Loon LC: NPR1: the spider in the web of induced resistance signaling pathways. *Curr Opin Plant Biol* 2004, 7(4):456-464.
27. Feys BJ, Parker JE: Interplay of signaling pathways in plant disease resistance. *Trends Genet* 2000, 16(10):449-455.

28. Despres C, DeLong C, Glaze S, Liu E, Fobert PR: The Arabidopsis NPR1/NIM1 protein enhances the DNA binding activity of a subgroup of the TGA family of bZIP transcription factors. *Plant Cell* 2000, 12(2):279-290.

29. Zhou JM, Trifa Y, Silva H, Pontier D, Lam E, Shah J, Klessig DF: NPR1 differentially interacts with members of the TGA/OBF family of transcription factors that bind an element of the *PR-1* gene required for induction by salicylic acid. *Mol Plant Microbe Interact* 2000, 13(2):191-202.

30. Ndamukong I, Abdallat AA, Thurow C, Fode B, Zander M, Weigel R, Gatz C: SA-inducible*Arabidopsis* glutaredoxin interacts with TGA factors and suppresses JA-responsive*PDF1.2* transcription. *Plant J* 2007, 50(1):128-139.

31. Johnson C, Mhatre A, Arias J: NPR1 preferentially binds to the DNA-inactive form of*Arabidopsis* TGA2. *Biochim Biophys Acta* 2008, 1779(10):583-589.

32. Fan W, Dong X: In vivo interaction between NPR1 and transcription factor TGA2 leads to salicylic acid-mediated gene activation in Arabidopsis. *Plant Cell* 2002, 14(6):1377-1389.

33. Boyle P, Le Su E, Rochon A, Shearer HL, Murmu J, Chu JY, Fobert PR, Despres C: The BTB/POZ domain of the Arabidopsis disease resistance protein NPR1 interacts with the repression domain of TGA2 to negate its function. *Plant Cell* 2009, 21(11):3700-3713.

34. Spoel SH, Koornneef A, Claessens SM, Korzelius JP, Van Pelt JA, Mueller MJ, Buchala AJ, Metraux JP, Brown R, Kazan K, *et al.*: NPR1 modulates cross-talk between salicylate- and jasmonate-dependent defense pathways through a novel function in the cytosol. *Plant Cell* 2003, 15(3):760-770

35. Leon-Reyes A, Spoel SH, De Lange ES, Abe H, Kobayashi M, Tsuda S, Millenaar FF, Welschen RA, Ritsema T, Pieterse CM: Ethylene modulates the role of NONEXPRESSOR OF PATHOGENESIS-RELATED GENES1 in cross talk between salicylate and jasmonate signaling. *Plant Physiol* 2009, 149(4):1797-1809.

36. Koornneef A, Pieterse CM: Cross talk in defense signaling. *Plant Physiol* 2008, 146(3):839-844.

37. Malnoy M, Jin Q, Borejsza-Wysocka EE, He SY, Aldwinckle HS: Overexpression of the apple *MpNPR1* gene confers increased

disease resistance in *Malus* x *domestica*. *Mol Plant Microbe Interact* 2007, 20(12):1568-1580.

38. Sakamoto K, Tada Y, Yokozeki Y, Akagi H, Hayashi N, Fujimura T, Ichikawa N: Chemical induction of disease resistance in rice is correlated with the expression of a gene encoding a nucleotide binding site and leucine-rich repeats. *Plant Mol Biol* 1999, 40(5):847-855.
39. Yang Y, Qi M, Mei C: Endogenous salicylic acid protects rice plants from oxidative damage caused by aging as well as biotic and abiotic stress. *Plant J* 2004, 40(6):909-919.
40. Chern M, Fitzgerald HA, Canlas PE, Navarre DA, Ronald PC: Overexpression of a rice*NPR1* homolog leads to constitutive activation of defense response and hypersensitivity to light. *Mol Plant Microbe Interact* 2005, 18(6):511-520.
41. Hebbar PK: Cacao diseases: a global perspective from an industry point of view. *Phytopathology* 2007, 97(12):1658-1663.
42. Purdy LH, Schmidt RA: STATUS OF CACAO WITCHES' BROOM: biology, epidemiology, and management. *Annu Rev Phytopathol* 1996, 34:573-594.
43. Wood G, Lass R: *Cocoa*. New York: Longman Scientific & Technical Press; 1985.
44. Evans H, Priori C: Cocoa pod diseases: causal agents and control. *Outlook on Agriculture* 1987, 16:35-41.
45. Phillips-Mora W, Wilkinson MJ: Frosty pod of cacao: a disease with a limited geographic range but unlimited potential for damage. *Phytopathology* 2007, 97(12):1644-1647.
46. Bailey BA, Bae H, Strem MD, Antunez de Mayolo G, Guiltinan MJ, Verica JA, Maximova SN, Bowers JH: Developmental expression of stress response genes in *Theobroma cacao*leaves and their response to Nep1 treatment and a compatible infection by*Phytophthora megakarya*. *Plant Physiol Biochem* 2005, 43(6):611-622.
47. Bailey BA, Strem MD, Bae H, de Mayolo GA, Guiltinan MJ: Gene expression in leaves of*Theobroma cacao* in response to mechanical wounding, ethylene, and/or methyl jasmonate. *Plant Science* 2005, 168(5):1247-1258.

48. Buddenhagen IW: Resistance and Vulnerability of Tropical Crops in Relation to Their Evolution and Breeding. *Annals of the New York Academy of Sciences* 1977, 287(Feb25):309-326.

49. Guiltinan M, Verica J, Zhang D, Figueira A: Genomics of *Theobroma cacao*, "The foods of the Gods". In *Genomics of Tropical Crop Plants*. Edited by Moore PM, Ming R. New York: Springer; 2008.

50. Schnell RJ, Kuhn DN, Brown JS, Olano CT, Phillips-Mora W, Amores FM, Motamayor JC:Development of a marker assisted selection program for cacao. *Phytopathology* 2007, 97(12):1664-1669.

51. Bae H, Kim SH, Kim MS, Sicher RC, Lary D, Strem MD, Natarajan S, Bailey BA: The drought response of *Theobroma cacao* (cacao) and the regulation of genes involved in polyamine biosynthesis by drought and other stresses. *Plant Physiol Biochem* 2008, 46(2):174-188.

52. Maximova SN, Marelli JP, Young A, Pishak S, Verica JA, Guiltinan MJ: Over-expression of a cacao class I chitinase gene in *Theobroma cacao L.* enhances resistance against the pathogen, *Colletotrichum gloeosporioides*. *Planta* 2006, 224(4):740-749.

53. Verica JA, Maximova SN, Strem MD, Carlson JE, Bailey BA, Guiltinan MJ: Isolation of ESTs from cacao (*Theobroma cacao L.*) leaves treated with inducers of the defense response. *Plant Cell Rep* 2004, 23(6):404-413.

54. Jones PG, Allaway D, Gilmour DM, Harris C, Rankin D, Retzel ER, Jones CA: Gene discovery and microarray analysis of cacao (*Theobroma cacao L.*) varieties. *Planta* 2002, 216(2):255-264.

55. Argout X, Fouet O, Wincker P, Gramacho K, Legavre T, Sabau X, Risterucci AM, Da Silva C, Cascardo J, Allegre M, *et al.*: Towards the understanding of the cocoa transcriptome: Production and analysis of an exhaustive dataset of ESTs of *Theobroma cacao L.* generated from various tissues and under various conditions. *BMC Genomics* 2008, 9:512.

56. Albagli O, Dhordain P, Deweindt C, Lecocq G, Leprince D: The BTB/POZ domain: a new protein-protein interaction motif common to DNA- and actin-binding proteins. *Cell Growth Differ* 1995, 6(9):1193-1198.

57. Li J, Mahajan A, Tsai MD: Ankyrin repeat: a unique motif mediating protein-protein interactions. *Biochemistry* 2006, 45(51):15168-15178.

58. Zhang Y, Fan W, Kinkema M, Li X, Dong X: Interaction of NPR1 with basic leucine zipper protein transcription factors that bind sequences required for salicylic acid induction of the *PR-1* gene. *Proc Natl Acad Sci USA* 1999, 96(11):6523-6528.

59. Kinkema M, Fan W, Dong X: Nuclear localization of NPR1 is required for activation of*PR* gene expression. *Plant Cell* 2000, 12(12):2339-2350.

60. Higo K, Ugawa Y, Iwamoto M, Korenaga T: Plant cis-acting regulatory DNA elements (PLACE) database: 1999. *Nucleic Acids Res* 1999, 27(1):297-300

61. Lescot M, Dehais P, Thijs G, Marchal K, Moreau Y, Van de Peer Y, Rouze P, Rombauts S:PlantCARE, a database of plant cis-acting regulatory elements and a portal to tools for in silico analysis of promoter sequences. *Nucleic Acids Res* 2002, 30(1):325-327.

62. Rombauts S, Dehais P, Van Montagu M, Rouze P: PlantCARE, a plant cis-acting regulatory element database. *Nucleic Acids Res* 1999, 27(1):295-296.

63. Molina C, Grotewold E: Genome wide analysis of *Arabidopsis* core promoters. *BMC Genomics* 2005, 6(1):25.

64. Yang C, Bolotin E, Jiang T, Sladek FM, Martinez E: Prevalence of the initiator over the TATA box in human and yeast genes and identification of DNA motifs enriched in human TATA-less core promoters. *Gene* 2007, 389(1):52-65.

65. Lebel E, Heifetz P, Thorne L, Uknes S, Ryals J, Ward E: Functional analysis of regulatory sequences controlling *PR-1* gene expression in Arabidopsis. *Plant J* 1998, 16(2):223-233.

66. Yu D, Chen C, Chen Z: Evidence for an important role of WRKY DNA binding proteins in the regulation of NPR1 gene expression. *Plant Cell* 2001, 13(7):1527-1540.

67. Yang P, Chen C, Wang Z, Fan B, Chen Z: A pathogen- and salicylic acid-induced WRKY DNA-binding activity recognizes the elicitor response element of the tobacco class I chitinase gene promoter. *Plant J* 1999, 18(2):141-149.

68. Faleiro F, Queiroz V, Lopes U, Guimarães C, Pires J, Yamada M, Araújo I, Pereira M, Schnell R, Filho G, *et al.*: Mapping QTLs for witches' broom (*Crinipellis Perniciosa*) resistance in cacao (*Theobroma Cacao L*). *Euphytica* 2006, 149(1):227-235.

69. Cao H, Glazebrook J, Clarke JD, Volko S, Dong X: The *Arabidopsis NPR1* gene that controls systemic acquired resistance encodes a novel protein containing ankyrin repeats. *Cell* 1997, 88(1):57-63.

70. Spoel SH, Mou Z, Tada Y, Spivey NW, Genschik P, Dong X: Proteasome-mediated turnover of the transcription coactivator NPR1 plays dual roles in regulating plant immunity. *Cell* 2009, 137(5):860-872.

71. Zhang X, Chen S, Mou Z: Nuclear localization of NPR1 is required for regulation of salicylate tolerance, *isochorismate synthase 1* expression and salicylate accumulation in *Arabidopsis*. *J Plant Physiol* 2010, 167(2):144-148.

72. Penninckx IA, Thomma BP, Buchala A, Metraux JP, Broekaert WF: Concomitant activation of jasmonate and ethylene response pathways is required for induction of a plant defensin gene in Arabidopsis. *Plant Cell* 1998, 10(12):2103-2113.

73. Endah R, Beyene G, Kiggundu A, van den Berg N, Schluter U, Kunert K, Chikwamba R:Elicitor and *Fusarium*-induced expression of *NPR1-like* genes in banana. *Plant Physiol Biochem* 2008, 46(11):1007-1014.

74. Zhang Y, Wang X, Cheng C, Gao Q, Liu J, Guo X: Molecular cloning and characterization of *GhNPR1*, a gene implicated in pathogen responses from cotton (*Gossypium hirsutum L*). *Biosci Rep* 2008, 28(1):7-14.

75. Irish VF, Yamamoto YT: Conservation of floral homeotic gene function between Arabidopsis and antirrhinum. *Plant Cell* 1995, 7(10):1635-1644.

76. Maizel A, Busch MA, Tanahashi T, Perkovic J, Kato M, Hasebe M, Weigel D: The floral regulator LEAFY evolves by substitutions in the DNA binding domain. *Science* 2005, 308(5719):260-263.

77. Dong X, Braun EL, Grotewold E: Functional conservation of plant secondary metabolic enzymes revealed by complementation of Arabidopsis flavonoid mutants with maize genes. *Plant Physiol* 2001, 127(1):46-57.

78. Wang D, Amornsiripanitch N, Dong X: A genomic approach to identify regulatory nodes in the transcriptional network of systemic acquired resistance in plants. *PLoS Pathog* 2006, 2(11):e123.

79. Johnson C, Boden E, Arias J: Salicylic acid and NPR1 induce the recruitment of trans-activating TGA factors to a defense gene promoter in Arabidopsis. *Plant Cell* 2003, 15(8):1846-1858.

80. Argout X, Salse J, Aury J, Guiltinan M, Droc G, Gouzy J, Allegre M, Chaparro C, Legavre T, Maximova S, *et al.*: The Genome of Theobroma cacao. *Nature Genetics* 2010, in press.

81. Ploetz RC: Cacao diseases: important threats to chocolate production worldwide. *Phytopathology* 2007, 97(12):1634-1639.

82. Thompson JD, Higgins DG, Gibson TJ: CLUSTAL W: improving the sensitivity of progressive multiple sequence alignment through sequence weighting, position-specific gap penalties and weight matrix choice. *Nucleic Acids Res* 1994, 22(22):4673-4680.

83. Melnick RL, Zidack NK, Bailey BA, Maximova SN, Guiltinan M, Backman PA: Bacterial endophytes: *Bacillus* spp. from annual crops as potential biological control agents of black pod rot of cacao. *Biological Control* 2008, 46(1):46-56.

84. Brechenmacher L, Weidmann S, van Tuinen D, Chatagnier O, Gianinazzi S, Franken P, Gianinazzi-Pearson V: Expression profiling of up-regulated plant and fungal genes in early and late stages of *Medicago truncatula-Glomus mosseae* interactions. *Mycorrhiza* 2004, 14(4):253-262.

85. Hajdukiewicz P, Svab Z, Maliga P: The small, versatile pPZP family of *Agrobacterium*binary vectors for plant transformation. *Plant Mol Biol* 1994, 25(6):989-994.

86. Mitsuhara I, Ugaki M, Hirochika H, Ohshima M, Murakami T, Gotoh Y, Katayose Y, Nakamura S, Honkura R, Nishimiya S, *et al.*: Efficient promoter cassettes for enhanced expression of foreign genes in dicotyledonous and monocotyledonous plants. *Plant Cell Physiol* 1996, 37(1):49-59.

87. Lin JJ: Electrotransformation of *Agrobacterium*. *Methods Mol Biol* 1995, 47:171-178.

88. Clough SJ, Bent AF: Floral dip: a simplified method for *Agrobacterium*-mediated transformation of *Arabidopsis thaliana*. *Plant J* 1998, 16(6):735-743.

89. Maximova S, Miller C, Antunez de Mayolo G, Pishak S, Young A, Guiltinan MJ: Stable transformation of *Theobroma cacao L.* and influence of matrix attachment regions on GFP expression. *Plant Cell Rep* 2003, 21(9):872-883.

90. Katagiri F, Thilmony R, He S: The *Arabidopsis thaliana-Pseudomonas syringae*interaction. In *The Arabidopsis Book*. The American Society of Plant Biologists; 2002.

Chapter 4

Mir171h Restricts Root Symbioses and Shows Like its TargetNSP2 a Complex Transcriptional Regulation in Medicago Truncatula

Vinzenz Hofferek[1], Amelie Mendrinna[1], Nicole Gaude[1], Franziska Krajinski[1] and Emanuel A Devers[1,2]

[1]Max-Planck-Institute of Molecular Plant Physiology, Am Mühlenberg 1, Potsdam, 14476, (OT) Golm, Germany

[2]Present address: Department of Biology, Swiss Federal Institute of Technology Zurich, Zürich, Switzerland

ABSTRACT

Background

Legumes have the unique capability to undergo root nodule and arbuscular mycorrhizal symbiosis. Both types of root endosymbiosis are regulated by *NSP2*, which is a target of microRNA171h (miR171h).

Although, recent data implies that miR171h specifically restricts arbuscular mycorrhizal symbiosis in the root elongation zone of *Medicago truncatula* roots, there is limited knowledge available about the spatio-temporal regulation of miR171h expression at different physiological and symbiotic conditions.

Results

We show that miR171h is functionally expressed from an unusual long primary transcript, previously predicted to encode two identical miR171h strands. Both miR171h and *NSP2* transcripts display a complex regulation pattern, which involves the symbiotic status and the fertilization regime of the plant. Quantitative Real-time PCR revealed that miR171h and *NSP2* transcript levels show a clear anti-correlation in all tested conditions except in mycorrhizal roots, where *NSP2* transcript levels were induced despite of an increased miR171h expression. This was also supported by a clear correlation of transcript levels of *NSP2* and *MtPt4*, a phosphate transporter specifically expressed in a functional AM symbiosis. MiR171h is strongly induced in plants growing in sufficient phosphate conditions, which we demonstrate to be independent of the *CRE1* signaling pathway and which is also not required for transcriptional induction of *NSP2* in mycorrhizal roots. *In situ* hybridization and promoter activity analysis of both genes confirmed the complex regulation involving the symbiotic status, P and N nutrition, where both genes show a mainly mutual exclusive expression pattern. Overexpression of miR171h in *M. truncatula* roots led to a reduction in mycorrhizal colonization and to a reduced nodulation by *Sinorhizobium meliloti*.

Conclusion

The spatio-temporal expression of miR171h and *NSP2* is tightly linked to the nutritional status of the plant and, together with the results from the overexpression analysis, points to an important function of miR171h to integrate the nutrient homeostasis in order to safeguard the expression domain of*NSP2* during both, arbuscular mycorrhizal and root nodule symbiosis.

BACKGROUND

Plants constantly have to cope with phosphate (P_i) limiting conditions and one strategy to overcome P_i limitation is the development of a mutualistic association called arbuscular mycorrhizal symbiosis (AMS), which is formed by most land plants and fungi of the phylum *Glomeromycota*. AMS can enhance phosphate uptake and growth of the plant [[1],[2]] and is named for the formation of intracellular tree like structures called arbuscules. Arbuscules are mostly formed in the inner cortical cell layer of mycorrhizal roots and are always surrounded by the plant-derived periarbuscular membrane (PAM), the site of nutrient exchange [[3],[4]]. The formation of AMS is initiated after a chemical dialogue between the host plant and symbiont [[5]]. The plant secretes strigolactones, a group of plant hormones known to stimulate of fungal spore germination and hyphal branching [[6]]. In return, the fungus releases a complex mixture of lipochito-oligosaccharides, called Myc-LCOs or Myc-factors, which stimulate the formation of AMS and induces lateral root formation in the legume plant *Medicago truncatula* [[7]].

Strigolactone synthesis and secretion is induced by P_i limitation [[8]–[11]]. Biosynthesis of strigolactones requires the GRAS type transcription factors N*odulation Signaling* Path*way* (NSP) *1* and*NSP2*, and *nsp1nsp2* double mutants show a reduced colonization by mycorrhizal fungi [[12]]. Interestingly, *NSP1* and *NSP2* are key components of the Nod-factor signaling pathway leading to root nodule formation [[13],[14]]. Recent results [[7]] suggest that *NSP2* is also involved in AMS. The authors showed that *nsp2* mutant plants do not respond to Myc-LCOs and are less colonized. Furthermore, NSP2 interacts with another GRAS transcription factor *R*equired for *A*rbuscular*M*ycorrhization (RAM) 1, and controls the expression of the glycerol-3-phosphate acyltransferase*RAM2*. Mutations in both genes lead to strongly reduced colonization by *Rhizophagus irregularis*. Interestingly, colonization with the pathogenic oomycete *Phytophthora palmivora* is also impaired in ram1 and ram2 mutants (if otherwise, need to specify) [[15],[16]]. The important role of *NSP2* in strigolactone biosynthesis, RNS and AMS implies that *NSP2* is an integral component of the common signaling pathway [[17]]. Therefore, it can be expected that the spatial and temporal expression of*NSP2* is tightly controlled.

Micro RNAs (miRNAs) are key regulators of gene expression and act by target transcript cleavage and/or translational repression [[18],[19]]. These small non-coding RNA molecules are predominantly 21 nt in size and have an important role in regulating developmental processes, hormonal signaling, organ polarity, RNA metabolism, and abiotic and biotic stresses of the plant [[20]–[24]]. Some miRNAs, e.g. miR166 and miR169, have also been found to be involved in root nodule symbiosis [[25]–[27]]. First evidence that miRNAs are also involved in AMS came from the observation that multiple miRNAs, e.g. miR399, are differentially regulated in the shoots and roots of mycorrhizal *M. truncatula,* tobacco and tomato plants [[28],[29]]. Cloning and deep sequencing of the small RNAs and the degradome of mycorrhizal *M. truncatula* roots identified many miRNAs and their target mRNAs, of which several were differentially expressed in mycorrhizal roots [[30]]. Interestingly, a novel member of the miRNA171 family, miR171h, was shown to target *NSP2* [[30],[31]]. Given the above-mentioned role of *NSP2*, miR171h has been implicated in regulating root endosymbiosis by controlling a key component of the Sym-pathway [[27]]. This assumption has been strengthened by showing miRNA171h expression affects the mycorrhizal colonization, is induced by Myc-LCOs, and is conserved among mycotrophic plants [[32]]. Also, it was shown that the expression of both, miR171h and *NSP2,* is induced upon cytokinin treatment and that this regulation is dependent on Cytokinin Response1 (CRE1) [[33]]. Cytokinins and *CRE1* are involved in nodule organogenesis [[34]], but cytokinins have also been implicated to be involved in arbuscular mycorrhizal symbiosis [[35]]. Additionally, a recent study employing deep sequencing of *Lotus japonicus* nodules revealed a non-canonical miR171 isoform, related to *Medicago* miR171h, which targets *LjNSP2* [[36]]. These results indicate an additional regulating role of miR171h in nodule symbiosis, however a direct involvement could not be demonstrated so far [[32],[36]].

In this study we show that miR171h negatively regulates both types of root endo-symbioses through perception of the nutritional status of the host plant and shows a mutually exclusive expression pattern with its target *NSP2* in the root cortex of *M. truncatula* plants.

RESULTS

Expression of an 811 bp Mir171h Primary Transcript Mediates NSP2 Transcript Cleavage in Vivo

To confirm the functionality of the predicted 811 bp primary transcript of miR171h [[31]] and the potential to silence *NSP2 in vivo*, we applied miRNA sensor constructs. Three different constructs were applied (Figure 1A). The first construct, MIR171h-GFP, contained an 811 bp fragment of the MIR171h primary transcript [[31]], which was constitutively expressed by a 35S-promoter. The construct includes an independent and constitutively expressed eGFPer as a visual transformation control. The second construct, *m*iR171h *b*inding *s*ite (MBS)-NSP2, represented the actual miRNA sensor. It was composed of a 35S-promoter driven mRFP fused to five repeats of the miR171h binding site of *NSP2*. As a control, the MBS of *NSP2* was mutated to a scrambled sequence (MBS-mut), which was unable to bind miR171h. The constructs were used in *Agrobacterium tumefaciens*-mediated tobacco leaf infiltration assays and the mRFP fluorescence around the infiltration site was monitored (Figure 1B). When MBS-NSP2 was co-infiltrated with MIR171H-GFP, the mRFP fluorescence was abolished. The fluorescence was restored when MBS-mut and MIR171H-GFP were co-infiltrated, indicating that the loss of mRFP fluorescence was specifically due to miR171h-mediated sensor cleavage. The loss of fluorescence was due to drastically reduced mRFP protein levels in MIR171H-GFP co-infiltrated tobacco leaves of MBS-NSP2 compared to MBS-mut and single infiltration (Figure 1C). These results confirmed that*NSP2* is regulated by miR171h through specific binding of this miRNA to its previously identified binding site and is consistent with previous degradome results [[30]] and RACE experiments [[32]].

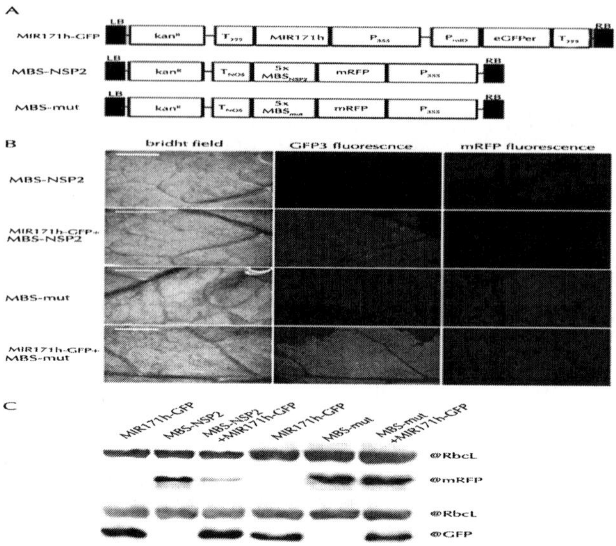

Figure 1: *In vivo*confirmation of*NSP2*gene silencing by miR171h using MIR171h overexpression and mRFP sensor constructs. (A) T-DNA structure of vectors used for leave infiltration experiments. MiR171h overexpression construct (MIR171h-GFP) in pK7WG2D [[37]] and sensor constructs with either wild-type (MBS-NSP2) or mutated miR171h (MBS-mut) binding site of NSP2 cloned in pGWB455 [[38]]. LB: left boarder, Kan^R: kanamycin resistance gene (nptII), Tnos: nopaline synthase terminator, MIR171h: miR171h primary transcript, P_{35S}: 35S promoter, green-fluorescent protein (GFP) cassette (pRolD–EgfpER–t35S), 5xMBS$_{NSP2}$: 5 repeats of miR171h binding site sequence of NSP2, 5xMBS$_{mut}$: 5 repeats of a mutated version of the miR171h binding site sequence of NSP2. (B) Co-infiltration of miR171h overexpression and mRFP sensor constructs.*Nicotiana benthamiana* leaves were infiltrated with the two sensor constructs MBS-NSP2 or MBS-mut. For each sensor construct, co-infiltration experiments with the MIR171h-GFP construct were carried out. Note the decreased mRFP fluorescence due to miR171h-mediated cleavage of mRFP sensor. Bright field images, GFP3 fluorescence and mRFP fluorescence are shown. Scale bar: 5 mm. (C) Western blot to prove miR171h cleavage of the miR171h binding site within the NSP2 sequence. Proteins were extracted from leaves infiltrated with MIR171h-GFP, MBS-NSP2 or MBS-mut and co-infiltration of both constructs. The upper part of the picture shows a western blot where mRFP was detected, indicating the presence of the sensor; the lower part shows a western blot with detection of GFP, indirectly indicating the presence of miR171h. On both blots, RuBisCO proteins were detected to demonstrate equal loading of the protein samples.

MiR171h and NSP2 Transcript Levels are affected by the Symbiotic Status of the Root and by P and N Levels

Previous studies suggested that miR171h is induced in the root elongation zone of mycorrhizal roots and that *NSP2* transcript levels are slightly repressed in mycorrhizal roots [[32]]. Also, miR171h transcription is directly induced by high phosphate nutrition [[30]]. To investigate the transcriptional regulation of *NSP2* and miR171h in more detail, we analyzed both transcript levels in response to mycorrhizal symbiosis and root nodulation, and under different phosphate and nitrogen fertilization treatments. For this purpose we compared the abundance of mature miR171h to the relative transcript abundance of *NSP2* after normalization to full nutrition (+P+N) conditions (Figure 2). As expected, miR171h accumulation was repressed by phosphate starvation ($P \leq 0.05$), i.e. positively influenced by high phosphate conditions. It also increases in nodulated roots, in which case it is further enhanced by nitrogen starvation. These results indicate that both *NSP2* and MIR171h show a complex regulation of their transcription, which depends on the symbiotic status of the roots and the nutrient fertilization regime.

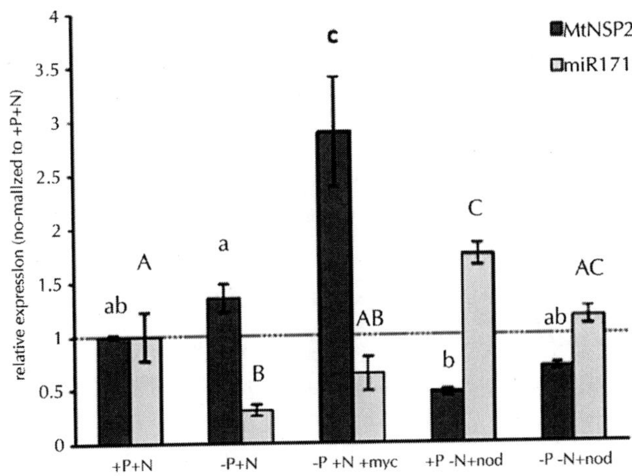

Figure 2: Relative expression levels of mature miR171h and*NSP2*transcripts in*M. truncatula*roots. -P: 20 μM phosphate, +P: 1 mM phos-

phate, −N: 0 mM; +N 5 mM nitrate fertilization, myc: mycorrhizal roots, nod: nodulated by *Sinorhizobium meliloti*. Plants were harvested 3 wpi. Normalization of the expression data was carried out against a reference gene index (MtPdf2 and MtEf1) and the resulting relative expression was normalized to full nutrition condition (+P+N). Data shown are average values of 3–4 biological with two technical replicates each. Error bars indicate the standard errors. Different letters indicate statistical different values ($P<0.05$, two-way ANOVA with Holm-Sidak multiple comparison).

NSP2 Transcript Levels in mycorrhizal Roots are elevated despite of Increased MiR171h Expression

A clear anti-correlation ($r=-0.98$; $p<0.05$) of miR171h and *NSP2* accumulation was present in all but the mycorrhizal condition (Figure 2). In the conditions used, *NSP2* shows the highest relative transcript abundance in mycorrhizal roots as compared to all other conditions tested.

To investigate the miR171h and *NSP2* expression in mycorrhizal roots over a time-course of AM symbiosis development, an experiment of 6 weeks was carried out and RNA accumulation of marker genes for AM symbiosis development and function were analyzed. This clearly showed that the expression of pri-miR171h increases in mycorrhizal roots from 2 weeks post inoculation on, as compared to non-mycorrhizal roots under the applied conditions (−P, +N). Additionally, the time-course confirms elevated *NSP2* transcript levels in mycorrhizal roots, despite of enhanced miR171h accumulation. Therefore we assume that the abundance of *NSP2* transcript levels in mycorrhizal roots is maintained by a miR171h-independent factor.

To further investigate the *NSP2* transcript levels in mycorrhizal roots, we analyzed if *NSP2* transcript level shows a correlation to *MtPt4* transcript levels in individual plants. *MtPt4* encodes a phosphate transporter specifically expressed in arbuscule-containing cells [[39]] and can be regarded as a marker for a functional AM symbiosis. We found a clear positive correlation ($r=0.927$, $P<0.01$) of*NSP2* and *MtPt4* transcript levels (Figure 3). This supports the assumption that *NSP2* is induced in mycorrhizal roots by a mycorrhiza-dependent factor.

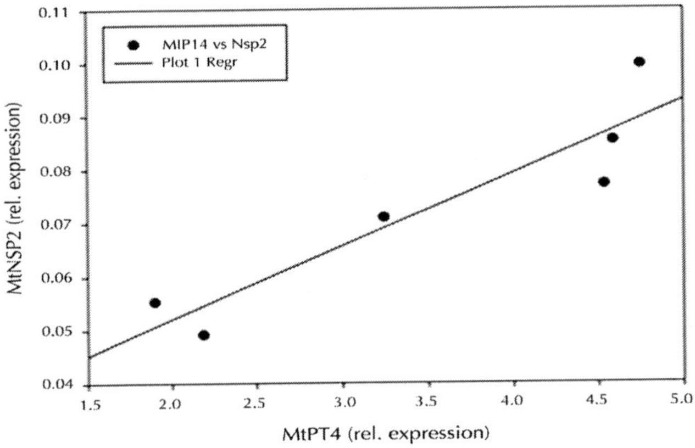

Figure 3: The relative transcript abundance of*NSP2*positively correlates with MtPt4. Scatter plot of the relative expression of *NSP2*against the relative expression of MtPt4 of individual WT plants including a linear regression (black line). A statistically significant correlation was calculated (r=0.927, P <0.01, Pearson product moment correlation). All plants were harvested 4 wpi. Normalization of the expression data was carried out against a reference gene index (*MtPdf2* and *MtEf1*).

The Suppression of Functional Symbiotic Structures by High Phosphate Fertilization Is Independent of NSP2

MiR171h transcript levels are increased in plants supplied with high phosphate concentrations. Additionally, at high phosphate conditions mycorrhizal colonization is often decreased as compared to plants grown at phosphate starvation and additionally results in less functional symbiotic structures, namely arbuscules [[28],[40]]. Therefore, is might be assumed that the suppression of mycorrhizal symbiosis at high phosphate conditions is dependent on the miR171h-*NSP2* regulon. To address this question, we inoculated wild type and *nsp2* mutant plants with *R. irregularis* growing in phosphate starvation (20 μM P_i) or high phosphate (1 mM P_i) conditions. Transcript levels of MtPT4 as a marker for a functional AM symbiosis were measured at three weeks after inoculation. Mycorrhizal wild-type plants and *nsp2* mutants showed significantly decreased levels of MtPT4 expression when grown at high phosphate fertilization, which indicates a significant lower frequency of functional symbiotic structures of both lines. Hence, suppression of mycorrhizal colonization by high phosphate fertilization is independent of *NSP2*.

Induced Expression of MiR171h at High Phosphate Conditions is Not Dependent on Cre1

A recent study of *cis* regulatory elements of the *NSP2* promoter gave evidence that *NSP2*, as well MIR171h, transcription is directly influenced by cytokinin and depends on the cytokinin receptor *CRE1*[[33]]. To analyze whether the induction of miR171h at high phosphate and *NSP2* induction in mycorrhizal roots is dependent on *CRE1*-mediated cytokinin perception, we investigated the relative expression levels of miR171h and *NSP2* in both conditions using *cre1-1* mutant plants and wild-type control plants.

The *cre1-1* mutation has subtle effect on the *NSP2* transcript abundance compared to wild-type plants in regard to the increased expression in mycorrhizal plants (Figure 4B). On the one hand, only the increase of the *NSP2* transcript abundance in wild-type plants is statistically significant ($P \leq 0.05$) compared to *cre1-1* plants, whereas on the other hand there is no statistical significance between wild-type and *cre1-1* plants at mycorrhizal conditions. This might indicate that the transcriptional induction of *NSP2* in mycorrhizal roots is not directly dependent of *CRE1*, however we cannot rule out that *NSP2* expression levels might be indirectly influenced by *CRE1* during arbuscular mycorrhizal symbiosis. We also investigated the effect of *cre1-1* on the AM symbiosis by monitoring the relative transcript levels of the mycorrhizal marker genes *MtPt4* and *RiTEF* [[3],[41]]. Expression level of both genes did not differ between mycorrhizal *cre1-1* and wild type plants (Figure 4C) at 4 weeks post inoculation (wpi), which indicates that *cre1-1* mutants are not impaired in mycorrhizal development. That is also supported by a similar correlation between *NSP2* and *MtPt4* relative transcript levels in*cre1-1* plants compared to the wild type. In contrast to the nodule symbiosis, where *CRE1* is clearly involved in miR171h and *NSP2* regulation [[33]], these results do not clearly demonstrate that the induction of *NSP2* in mycorrhizal roots is directly mediated by *CRE1*.

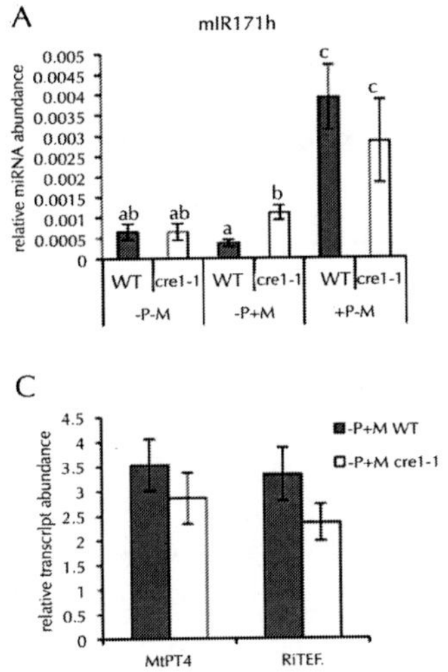

Figure 4: The*cre1-1*mutation does not affect phosphate dependent regulation of miR171h as well as*NSP2*transcript abundance and has no effect on mycorrhizal induced expression of*NSP2*and arbuscular mycorrhizal marker genes. The relative expression of mature miR171h(A) or *NSP2*(B) in roots of either WT plants (grey bars) or cre1-1 plants (white bars) determined at different phosphate fertilization and mycorrhizal conditions. -P: 20 μM phosphate, +P: 1 mM phosphate, −M: non-mycorrhizal roots, +M: mycorrhizal roots. (C) The relative expression of mycorrhizal the marker genes *MtPt4* and *RiTEF*were determined in root material from the same plants shown in A and B (−P+M). All plants were harvested 4 wpi. Normalization of the expression data was carried out against a reference gene index (*MtPdf2* and *MtEf1*). Data shown are average values of 3–6 biological replicates. Error bars indicate the standard errors. Different letters indicate statistical different values ($P<0.05$). A and B: two-way ANOVA with Holm-Sidak multiple comparison. C: one-way ANOVA).

Both *cre1-1* and wild type plants showed a significant increase in the amount of miR171h at high $P_i$conditions (1 mM) compared to either non-mycorrhizal or mycorrhizal low P_i conditions (20 μM) (Figure 4A). At high phosphate conditions, no significant difference in the relative abundance of miR171h between *cre1-1* and wild type plants could be observed, indicating that the phosphate-induced expression of miR171h is not mediated by *CRE1*.

Reporter Fusions Confirm the Complex and Spatial Regulation of NSP2- And MiR171h-Promoter Activities

An increased level of both NSP2 and miR171h in mycorrhizal roots suggests that these transcripts might be spatially separated in roots. We therefore used promoter-reporter fusions to localize the promoter activity of MIR171h and *NSP2* in roots grown at different nutritional and symbiotic conditions. A 1248 bp fragment upstream of the *NSP2* coding sequence and a 900 bp fragment upstream of the miR171h primary transcript were fused to a β-glucuronidase (GUS) reporter gene. Both promoter-reporter constructs were transformed into *M. truncatula* roots by *Agrobacterium rhizogenes*-mediated transformation.

As expected from the transcript accumulation pattern, promoter activity pattern of both MIR171h and*NSP2* showed drastic changes in response to different phosphate and nitrate fertilization treatments (Figure 5). At phosphate starvation, the promoter of MIR171h was mainly active in the central cylinder and the endodermis (Figure 5B). However, after colonization with *R. irregularis*, a weak GUS signal was detectable in distinct, but not in all, arbuscule-containing cells (Figure 5A″). The *NSP2*promoter also showed a strong activity in the central cylinder and endodermis during phosphate starvation. In addition, the *NSP2* promoter showed a weak activity in cortical cells of non-mycorrhizal roots (Figure 5F), whereas in mycorrhizal roots an additional strong activity was observed in the epidermal cell layer, root hairs and in some arbuscule-containing cells of the root (Figure 5E and E″).

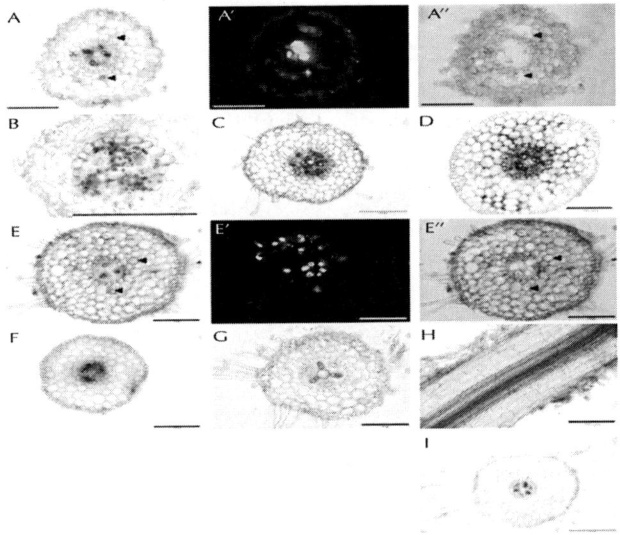

Figure 5: The*NSP2*and the miR171h promoter show distinct regulation in response to nutrients and mycorrhizal colonization.Roots were transformed with promoter-uidA fusions of either the MIR171h(A-D) or the *NSP2*(E-I) promoter. After root transformation, chimeric plants were potted into substrate and grown under different nutritional regimes: phosphate starvation (B and F), phosphate starvation and mycorrhizal colonization (A and E), nitrogen starvation (C, G, and H) or full nutrition (D and I). A, A′ and A″ as well as E, E′ and E′ represent identical root sections, with A′/E′ showing WGA Alexafluor 488 fluorescence to visualize *R. irregularis* structures and A″/E″ showing overlay of bright filed images and WGA Alexafluor 488 fluorescence. Scale bars represent 100 μm (black and white) or 200 μm (blue). Arrowheads point to GUS staining in cortical cells containing arbuscules.

This observation might be a reasonable explanation for the above mentioned increased transcript abundance of *NSP2* in mycorrhizal roots (Figure 2). During nitrogen starvation no major MIR171h-promoter activity could be observed except in some isolated cortical cells (Figure 5C), which might represent spontaneous promoter activity or staining artifacts. Increased miR171h expression seen during RNS (Figure 2) therefore is not due to the lack of nitrogen fertilization. In contrast to miR171h, the *NSP2* promoter showed a more complex activity pattern in roots of nitrogen-starved plants. Parts of the roots showed *NSP2* promoter activity in apparently random patterns in the cortex cells and endodermis (Figure 5G), whereas in different parts of the same root the promoter activity was observed in the whole cortex (Figure 5H). At full nutrition condition the MIR171h (Figure 5D) and *NSP2* promoter (Figure 5I) showed contrasting localizations, where the MIR171h promoter was active in all root cells with the strongest signals in the central cylinder. In turn, the *NSP2* promoter was not active in the root epidermis and cortex but active in the central cylinder and endodermis.

Next we analyzed the promoter activity of MIR171h and *NSP2* in root nodules of plants grown at high phosphate, without nitrate and inoculated with *Sinorhizobium meliloti*. Representative young and mature nodules are shown in Figure 6. The MIR171h promoter showed only weak activity in young nodules (Figure 6A) but in mature nodules an increased activity was visible at the nodule tip with the strongest activity matching the meristematic zone and getting gradually weaker in the direction of the subsequent infection and nitrogen fixation zones (Figure 6B). It is worth mentioning that the root cortex cells flanking the nodules showed strong GUS activity, which is in contrast to non-inoculated control roots where no promoter activity was observed in the root cortex (Figure 5C).

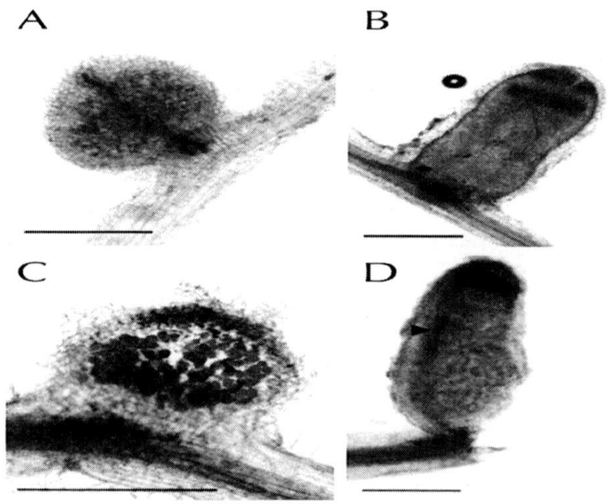

Figure 6: The*NSP2*and the MIR171h promoter show distinct expression domains in root nodules. Roots were transformed with promoter-

uidA fusions of either the MIR171h (A and B) or the NSP2 (C and D) promoter. After root transformation, chimeric plants were potted into substrate and grown under nitrogen starvation additionally colonized with *S. meliloti*. Pictures show either developing nodules (A and C) or mature nodules (B and D) from the same plants harvested at 5 wpi. Scale bars represent 500 µm. Arrowhead points to a vascular bundle.

The *NSP2* promoter was active in young and mature nodules with the strongest activity in the tip of young nodules and the vascular bundles of mature nodules (Figure 6C and D). Similar to the MIR171h promoter, the *NSP2* promoter was also active in the root cortex in the vicinity of young and mature nodules.

These results demonstrate that the promoter activities of *NSP2* and MIR171h are mainly mutually exclusive and are affected by phosphate as well as nitrate fertilization and by root colonization with *R. irregularis* and *S. meliloti*.

Mature MiR171h Accumulates the Meristematic Zone of Root Nodules

To localize the site of mature miR171h accumulation in mycorrhizal and nodulated roots, we used *in situ* hybridization with a miR171h-specific LNA probe (Figure 7). An LNA probe with a scrambled nucleotide sequence served as a negative control. Consistent with the location of the miR171h promoter activity, we found an accumulation of mature miR171h in the central cylinder of mycorrhizal roots (Figure 7A). We found only weak signals of mature miR171h accumulation in all arbuscule-containing cells, however the MIR171h promoter activity was only observed in distinct arbuscule-containing cells (Figure 5A). Based on the low signal to background ration it is not possible to draw a clear conclusion of its localization but its likely to be in arbuscule containing cells. Localization of the mature miRNA in root nodules was in agreement with the observed promoter activity pattern in mature nodules. Mature miR171h molecules accumulated to the meristematic zone of nodules with the signal intensity gradually decreasing towards the infection zone. No miR171h accumulation could be detected in the nitrogen fixation zone (Figure 7C). These results show that mature miR171h accumulation is consistent with its spatial origin of transcription and point to a protective role of miR171h against target mis-expression in root nodules and probably arbuscule containing cells.

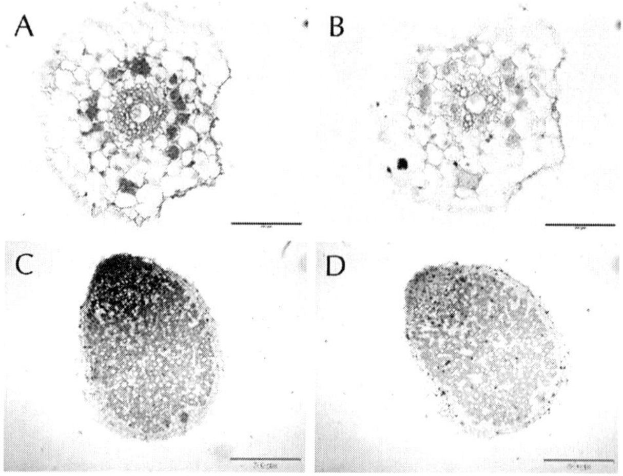

Figure 7: Localization of mature miR171h accumulation in mycorrhizal roots and root nodules via*in-situ*hybridization. Specific miR171h LNA probes (A and C) or scramble probes (B and D) were used for in situ hybridization to localize mature miR171h molecules in mycorrhizal roots (A,B) or root nodules (C,D). Detection was done using NBT/BCIP solution. Scale bars represent 100 μm (black) and 500 μm (blue).

Over-Expression of MiR171h Leads to a Reduced mycorrhizal Colonization and Reduced Nodule Numbers

To analyze the biological function of the observed spatial regulation of *NSP2* and miR171h transcription, we ectopically over-expressed miR171h in *M. truncatula* roots. For this purpose, *M. truncatula* roots were transformed with the construct, MIR171h-GFP (Figure 1), already used for the leaf infiltration assay to drive the constitutive expression of pri-miR171h. The empty vector was used as a control. The root-transformed plants were either inoculated with *R. irregularis* or *S. meliloti* to analyze the impact of miR171h over-expression on AMS or RNS, respectively. At 5 weeks post inoculation, over-expression of pri-miR171h led to a very strong accumulation of mature miR171h resulting in a drastic reduction of *NSP2* transcript levels (Figure 8A and B). In addition, we also observed a reduction in *MtDWARF27* transcript abundance, which is a downstream target of *NSP2* and important for converting the strigolactone orobanchol into didehydro-orobanchol [[12]]. Interestingly, we found significantly reduced intraradicular rRNA level of *R. irregularis* but not of *MtPt4* transcript level, which is similar to the unchanged *MtPt4* transcript levels in mycorrhizal roots of *nsp2-2* mutant plants. The microscopic analysis of the mycorrhizal phenotype [[42]] showed a concomitant reduction in the mycorrhizal intensity to similar levels as the *nsp2-2* mutant line (Figure 8A), but no significant reduction of the arbuscule abundance was observed in the pri-miR171h overexpressing plants (: Figure S6). This implies that plant roots which over-accumulated mature miR171h were less colonized by the mycorrhizal fungus whereas arbuscule development in colonized root areas was not impaired. Additionally, we also found significantly reduced nodule numbers in roots over-expressing mature miR171h compared to the vector controls (Figure 8B).

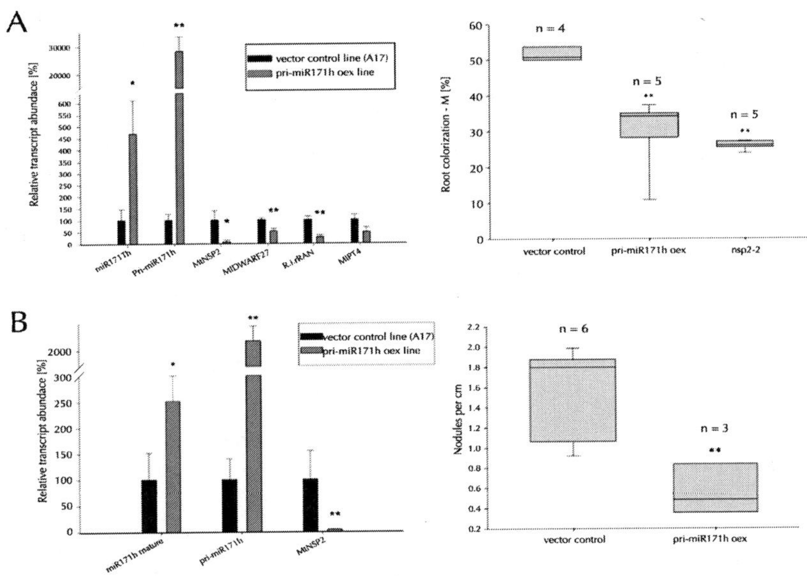

Figure 8: Ectopic overexpression of MIR171h leads to a strong repression of arbuscular mycorrhizal and root nodule symbiosis.Roots were transformed either the MIR171h-GFP (red) or the empty vector (black).

After root transformation, chimeric plants were potted into substrate and grown under different nutritional and symbiotic regimes: (A)20 μM phosphate inoculated with *R. irregularis*, (B) 0 M nitrate and inoculated with *S. meliloti*. Left panes show relative expression values (normalized to *MtPdf2* and*MtEf1*) of selected transcripts and mature miR171h and right panes display box-plots of symbiotic parameters, i.e. (A) mycorrhizal root colonization intensity [[42]] and (B) number of root nodules per cm of primary root length. Error bars indicate standard errors. Statistical different values between overexpression and vector control plants are indicated by single asterix ($P<0.05$) or double asterix ($P<0.01$) calculated by student's t-test. The number of biological replicates is indicated above box plots.

In summary, ectopic overexpression of the primary miR171h transcript in roots of *M. truncatula* is leading to a phenotype analogous to the *nsp2-2* mutant phenotype during mycorrhizal and nodule symbiosis.

DISCUSSION

One of the recently identified miRNAs, which specifically target genes essential for root endosymbiosis is mtr-miR171h [[30]].

MiR171h has later been predicted to be produced from an 811 bp long primary transcript encoding two miR171h loci on a single arm of its fold-back structure [[31]]. In this work, we provided experimental proof that miR171h is functionally expressed from an at least 811 bp long primary transcript. Cleavage of *NSP2* transcripts by miR171h has been verified by RACE-based methods [[30],[32],[36]]. However, so far functional analysis of miR171h has been carried out with the primarily identified partial precursor encoding only a single miRNA duplex [[32]]. By using a specific miRNA-sensor construct in a *N. benthamiana* leaf infiltration assay we could confirm *in-vivo* that the long double duplex-containing MIR171h primary transcript exhibits biological activity. This suggests that the long primary transcript described by Branscheid *et al.* [[28]] resembles the endogenous source of mature miR171h in *M. truncatula*.

The target of miR171h, *NSP2*, is a GRAS-type transcription factor, which is essential for the formation of root nodule symbiosis

[[13],[43],[44]], strigolactone biosynthesis [[12]] and is involved in the Myc-factor signaling pathway [[7]]. It has been recently shown that miR171h overexpression leads to a reduction of mycorrhizal root colonization [[32]] and it has been proposed to act specifically in arbuscular mycorrhizal symbiosis. Expression data of miR171h implies that this miRNA is induced by full nutrition conditions [[30]] and we hypothesized that miR171h is regulated by the phosphate status of the plant. Data from Arabidopsis and Rapeseed did not imply that the miR171 family is regulated by phosphate and nitrogen availability [[45]]. However, the data from the current study give strong evidence that the expression of miR171h is mainly induced by high phosphate availability (Figure 2). It is possible that the phosphate dependent regulation of a miR171 family member is dependent of the ability to undergo root endosymbiosis, because the miR171h isoform is only present in AMS capable plants [[27],[32]]. Whether additional miR171 family members are phosphate responsive in these plant species is not known. Interestingly, the promoter GUS reporter assays suggest that the expression of miR171h is not merely regulated by phosphate *per se* but rather by a combination of phosphate and nitrogen levels, because the activity of the miR171h promoter is absent at high phosphate and low nitrogen conditions compared to full nutrition conditions (Figure 5C and D).

Recently, it has been shown that miR171h expression is induced by cytokinin treatment and dependent on the *CRE1* pathway [[33],[46]] i.e. that the loss of *CRE1* leads to loss of miR171h induction by cytokinins. Because it is known that P_i and nitrogen fertilization increases cytokinin levels of plants [[47],[48]] and elevated levels are also found in shoots and roots of mycorrhizal plants [reviewed in 35]. Based on this knowledge we hypothesized that the expression levels of miR171h at high phosphate conditions might be dependent on *CRE1* as a direct effect of induced cytokinin levels in these plants. We used the *cre1-1* mutant plant line to address this question. The presented results indicate that phosphate dependent regulation of miR171h and its modest induction in mycorrhizal roots is not dependent on *CRE1*; the latter assumption is in agreement with the fact that the *cre1-1*mutation has no statistically significant effect on the mycorrhizal marker gene expression and thus mycorrhizal colonization (Figure 4C). Taken together we conclude that the miR171h expression is regulated by the phosphate status of the plant in a nitrogen dependent manner controlled by an unknown genetic factor, in addition to the previously demonstrated inducing effect of myc-LCOs [[32]].

Interestingly, we found elevated transcript levels of *NSP2* in mycorrhizal roots (Figure 2) between two and six weeks post inoculation. Similar to miR171h, the expression of *NSP2* is induced upon cytokinin treatment in a *CRE1*-dependent manner during high phosphate and low nitrogen conditions [[33]]. However, the involvement of *CRE1* in the induction of*NSP2* transcript abundance in mycorrhizal roots during low phosphate and high nitrogen conditions was not as clear as with the cytokinin treatment mentioned above (Figure 4B). Therefore, we assume that this up-regulation is mediated by an unknown factor and does not involve cytokinin perception*per se*. We cannot rule out a subtle influence *CRE1* on *NSP2* expression during arbuscular mycorrhizal symbiosis, though mycorrhizal colonization itself is clearly not dependent on the *CRE1*-pathway. The observed increase in *NSP2* transcript levels is most likely the result of the observed additional promoter activity in the epidermal cell layer of mycorrhizal roots compared to non-mycorrhizal roots (Figure 5E and F) and which lack MIR171h promoter activities (Figure 5A). We cannot rule out the possibility that this phenomenon is due to the lack of proper loading of Argonaute proteins, the major component of the RNA induced silencing complex (RISC), with mature miR171h, but due to a lack of antibodies against *M. truncatula* AGO1 homologues, this assumption awaits to be tested. Additionally,*NSP2* might be preferably repressed by a translational rather than post-transcriptional repression [[19]], which would leave the steady-state transcript abundance unchanged. However, owing to the mutually exclusive expression pattern of the miR171h-*NSP2* target pair during mycorrhizal symbiosis as discussed below, we favor the aforementioned explanation. Yet, our observation of an induced*NSP2* expression during mycorrhizal conditions is in contrast to the previously published results of Lauressergues *et al.* [[32]] who observed a slight decrease in *NSP2* transcript levels in mycorrhizal roots. The exact cause for this discrepancy is not known, but we could show that our results are consistent over 3 independent experiments: (I) by qRT-PCR measurement of *NSP2* transcript levels based on material 3 weeks post inoculation as shown in Figure 2, (II) by qRT-PCR measurement based on material from a time-course over 6 weeks as presented in, and (III) by qRT-PCR measurement based on the parental line of the *cre1-1* mutant, harboring a WT gene, 4 weeks post inoculation as shown in Figure 4. Moreover a correlation of single data points derived from individual plants based on the measurement shown in Figure 4 shows a positive correlation between *MtPt4* (as a measure of arbuscule abundance) and *NSP2*. Additionally, we reinvestigated publically available microarray datasets, which are deposited in the Medicago Gene Expression Atlas v3 (MtGEA; http://mtgea.noble.org *webcite*) using the probe Mtr.44789.1.S1_at, which corresponds to *NSP2*. Although the data of Gomez *et al.* [[49]] shows only a slight increase of NSP2 in mycorrhizal conditions at best, but no reduction as well, the data of Hogekamp *et al.* [[50]] shows a strong increase in the signal of the probe in response to 6 weeks of colonization by *R. irregularis* or *Glomus mosseae* and a strong decrease in response to 2 mM Pi treatment compared to 20 μM treated control conditions. Analysis of the supplementary data from Hogekamp *et al.* [[50]] in more detail revealed that the increase of *NSP2* expression in *R. irregularis* treated roots has a log2 fold change of 0.7 (p-value=0.02) and in *G. mosseae* 0.84 (p-value=0.008), which represents a roughly 1.7× times induction of *NSP2* expression during AM symbiosis, which is not only in agreement with our data and strongly supports our results, but also indicates that this response is conserved in symbiotic interactions with *G. mosseae*. As the slight induction in the data of Gomez *et al.* [[49]] is not significant and Lauressergues *et al.* [[32]] observed a reduction in *NSP2* levels upon*R. irregularis* colonization it is clear that there is at least one other unknown factor involved in the regulation of the miR171h-NSP2 target pair. These might involve specific regulation by additional miR171h target genes [[30],[32]] not investigated in this study, which might be preferably silenced by miR171h instead of *NSP2* at the conditions tested by us. Future detailed analysis of their spatial expression pattern from promoter to the protein level, the latter also accounts for *NSP2*, might give insight into this conundrum.

When comparing the spatial distribution of the MIR171h- and *NSP2*-promoter, it becomes evident that they predominantly show a mutually exclusive expression pattern, especially at full nutrition (cortex and root epidermis) and mycorrhizal conditions (root epidermis). Such a pattern has also been described for the miR395-SULTR2;1 miRNA target pair [[51]] and most likely functions to prevent expression of transcripts resulting from spontaneous promoter activity in specific tissues and cell types, a mechanism called promoter dampening [[18]]. That this model applies to the miR171h-*NSP2* target pair is in agreement with the results from the ectopic mis-expression of miR171h by 35S driven expression which leads to a drastic reduction in *NSP2* transcript levels and mycorrhizal colonization (this study and [[32]]) as well as with the 35S driven ectopic mis-expression of a miR171h-resistant NSP2 version leading to over-colonization of the root system and root apices, whereas ectopic mis-expression of the wild type *NSP2* transcript, which can be cleaved by miR171h, does not [[32]]. This mutually exclusive expression pattern is especially pronounced at full nutrition conditions where the miR171h is strongly expressed in all root tissues, whereas the *NSP2* promoter is completely inactive with the exception of the central cylinder. This assumes that the plant needs to comprehensively ensure a tight *NSP2* expression control in roots of nutrient replete plants to avoid the formation of root nodule and AM symbiosis, especially at full nutrition supply. Even though we cannot exclude the possibility that miR171h induction by high phosphate fertilization has an impact on mycorrhizal repression by restricting *NSP2* expression, we rule out that this effect is solely dependent on miR171h, because *nsp2-2* plants still respond to high phosphate treatment with repression of mycorrhizal colonization (: Figure S4) even though the *NSP2* dependent regulation is genetically knocked out, indicating that other regulatory circuits act epistatically to the miR171h-*NSP2* regulatory circuit. Additionally, a recent important discovery is that NSP2 interacts with another GRAS-type transcription factor RAM1, which is essential for mycorrhizal colonization and hyphopodia formation by controlling the expression of *RAM2* [[15],[16]]. RAM2, a glycerol-3-phosphate acyl transferase, catalyzes an important step in cutin monomer production, which seems to be an important molecule for promoting hyphopodia attachment to the plant root, and thus can also be hijacked by filamentous pathogenic fungi, e.g. *Phytophthora palmivora* [[16]]. Accordingly, it might be assumed that the mutually exclusive expression pattern of miR171h has been evolved to avoid hijacking of the *NSP2* signaling pathway by pathogens, creating only a narrow conditional window to allow the formation of beneficial symbiotic interactions, presumably relying on a previous signal from a beneficial symbiotic partner. We therefore conclude that miR171h acts as a promoter dampener, which protects specific cell-types of the root at distinct nutritional and symbiotic conditions from *NSP2* mis-expression in a mutually exclusive expression pattern, whereas the overall transcript abundance of *NSP2* in wild-type *Medicago truncatula* roots colonized by arbuscular mycorrhizal fungi is mainly determined by its promoter activity rather than direct cleavage by miR171h.

Other miRNA regulatory circuits often seen by miRNA-target pairs beside promoter dampening are spatial restriction and temporal regulation [[18]]. An exemplary spatial restriction type important in regulating root nodule development has been documented for miR169 in *M. truncatula* [[25]]. There, miR169 restricts the expression of its target *MtHAP2-1* to the meristematic zone of the root nodule, to allow correct maturation of the root nodule. The data of our study suggests that spatial restriction seems to be a predominant regulatory mechanism of miR171h during root nodule development, in contrast to the segregated expression pattern discussed above. In root nodules the MIR171h and*NSP2* promoter show an overlapping spatial distribution (Figure 6), with the *NSP2* promoter being active in the whole nodule. MIR171h transcription and accumulation is restricted to the meristematic zone, which was confirmed by *in-situ* hybridization (Figure 7). Ectopic over-expression of the long miR171h primary transcript led to a significant reduction in the number of root nodules (Figure 8) and therefore could demonstrate that miR171h over-expression represses root nodule symbiosis, additional to its described capability to restrict mycorrhizal root colonization [[32]]. Furthermore, a recent study from De Luis *et al.* [[36]] in *L. japonicus* identified a homologue of miR171h, lja-miR171c, in determinate root nodules, which similarly targets *LjNSP2*. Its expression in nodules is dependent on the presence of *Mesorhizobium loti*, which indicates that the regulation of miR171h expression has a higher level of complexity than previously thought. Previous attempts to overexpress miR171h and miR171c in *M. truncatula* and *L. japonicus* failed, which could be explained by the use of the 35S promoter, which seems to be diminished in arbuscule containing cells [[52]] and root nodules [[53],[54]] and thus might lead to a variation of the results. Our demonstrated phenotype might be explained by a dosage dependent effect of the miR171h overexpression, because the primary transcript used in this study encodes two miR171h duplexes [[31]] and thus the maximum abundance has likely been doubled, balancing the lower 35S activity in root nodules. Taken together, our data provides profound evidence that miR171h can directly influence the root nodule symbiosis and we hypothesize that miR171h protects the root of the plant from *NSP2* mis-expression in the meristematic zone of indeterminate nodules and therefore prevents bacterial invasion of this tissue, analogous to the protection of the root tip from colonization with mycorrhizal fungi [[32]].

CONCLUSIONS

Our data revealed that the spatio-temporal expression of miR171h and *NSP2* is tightly linked to the nutritional status of the plant and reflecting the different physiological conditions at which both types of endosymbiosis are favored by a host legume plant. Together with the results from the overexpression analysis, our data points to an important function of miR171h to integrate the nutrient homeostasis in order to safeguard the expression of *NSP2* during both, arbuscular mycorrhizal and root nodule symbiosis.

METHODS

Plant Materials and Growth Conditions

Seeds of *Medicago truncatula* cv. Jemalong (A17), *nsp2-2* and *cre1-1* mutant plants were germinated as described in [[28]]. Seedlings were grown in a quartz sand (0.6 - 1.2 mm), vermiculite and expanded clay substrate (7:1:1 [v/v]) with 16 h light (25°C) and were twice a week fertilized, with half a strengths Hoagland solution containing 20 μM Pi. For high Pi conditions the solution contained 1 mM Pi and for experiments under N limitation $Ca(NO_3)_2$ and KNO_3 were substituted by $CaCl_2$ and KCl, respectively. For mycorrhizal colonization the seedlings were inoculated with *R. irregularis* by mixing the substrate with 10% (v/v) substrate from *Allium porrum* plants, which were used as host plants to propagate *R. irregularis* for at least 2 months. Plants used for analysis were harvested at time points indicated in the results section and representative parts of the roots were stained with WGA-Alexa Fluor 488 (Invitrogen) to estimate mycorrhizal colonization. Remaining roots were immediately frozen in liquid nitrogen and immediately used for total RNA extraction or stored at −80°C.

Molecular Cloning

The 811 bp fragment of pri-miR171h located at the genomic position 31,405,065..31,404,255 of chromosome 3 (Mt3.5) [[31]] was cloned into the *Cauliflower mosaic* virus 35S (short 35S) driven binary vector pK7WG2D [[37]]. The resulting construct was named MIR171h-GFP.

The miR171h binding site of MtNSP2 and a miR171h non-cleavable sequence (scramble) were obtained by gene synthesis by Eurofins® (MWG® Operon) and cloned into the 35S driven binary vector pGWB455 [[55]] used for N-terminal mRFP protein fusions, resulting in the constructs MBS-NSP2 or MBS-mut, respectively. Both sequences contained stop codons in frame with mRFP.

Promoter regions of mtr-miR171h and MtNSP2, 900 bp as well as 1248 bp upstream of the start codon, respectively, were amplified from wild type (*Medicago truncatula* cv. Jemalong) genomic DNA and cloned into the binary vector pKGWFS7 containing GFP:GUS fusion protein driven by the inserted vector.

Root Transformation and Leaf Infiltration Assays

Prior to root transformation seeds were surface sterilized (12 min conc. H_2SO_4, 5 min 6% NaClO) and germinated (2 days dark, 4°C, 2 days dark, 25°C) on 0.8% [w/v] water agar.

Roots of *Medicago* seedlings were transformed with *Agrobacterium rhizogenes* (ARqua1) as published by [[56]]. In short, roots were cut with a sterile razor blade and wounded site was dipped on a two day old *A. rhizogenes* culture, transferred to 0.8% [w/v] water agar in the dark for two days before transferring to selective Fahraeus medium containing 25 µg/ml of kanamycin and cultivated for 3 weeks in the phytotron (12 h/12 h light/dark cycle).

Tobacco leaf infiltration essay was conducted using 4 weeks old *Nicotiana benthamiana* plants prior to the flowering phase. All experiments were performed twice in three independent plants. Tobacco plants were saturated with water 2–4 h before infiltration. *Agrobacterium tumefaciens* (GV2260) containing the appropriate constructs was inoculated in 20 ml YEB medium containing the appropriate antibiotics and grown for two days at 28°C and 250 rpm in the dark. Cultures were pelleted in 50 ml falcon tubes by centrifugation for 15 min at 4000 rpm (Eppendorf, 5804), re-suspended in ice cold AS medium containing 10 mM MES, 10 mM $MgCl_2$, 150 µM Acetosyringon pH 5.6 and diluted to an OD_{600} of 0.8 to 1 with AS medium and further incubated for 3 h at RT under soft shaking (~50 rpm). Infiltration was carried out as co-infiltration essay were Agrobacteria cultures containing different constructs were mixed 1:1 prior to infiltration. For single construct infiltration event, Agrobacteria, containing the respective construct, were mixed 1:1 with empty Agrobacteria. Each leaf was infiltrated with approximately 500 µl into the abaxial side of the tobacco leaf according to [[57]]. Plants were grown under moderate light conditions for another 3 days before checked for fluorescence of the reporter proteins.

Inoculation with *Sinorhizobium Meliloti*

Inoculation of *M. truncatula* roots with *S. meliloti* strain 1021 was carried out as described in [[30]].

RNA Extraction

The RNA was extracted using the miRVANA miRNA isolation kit (Ambion) with a previous Plant Isolation Aid step (Ambion) according to the manufacturer's protocol.

RT- and qRT-PCR Analysis

Quantitative RT-PCR analysis was carried out as described previously [[30]]. In brief, qRT-PCR analysis was performed in a 10 µl reaction volume. cDNA was diluted 1:10 and 1 µl was used for the reaction. The mixture contained 4 µl of 0.5 µM primer-pair mix and 5 µl Maxima™ SYBR Green/ROX qPCR Master Mix (Fermentas). The qRT-PCR machine, ABI Prism 7900 HT (Applied Biosystems), was used with the following reaction set up: first step 95°C for 10 min, 40 cycles 95°C for 15 sec and 60°C for 60 sec, alternately. Finally, a dissociation stage (95°C for 15 sec, 60°C for 15 sec heating with ramp rate of 2% up to 95°C for 15 sec) for melting curve analysis was included. Data was collected and analyzed using SDS 2.4 (Applied Biosystems) and SigmaPlot v.12 (Systat). The threshold for Ct-values was set to 0.2 with an automatic baseline correction for all experiments. The averages of the Ct-values of *MtEF1-α*, *MtPDF2* and *MtGAPDH* were used to calculate the housekeeping gene-index (HK-index) for normalization of transcript abundance. Functional symbiotic structures where determined by measuring the *MtPt4* transcript abundance [[3]] and the relative colonization level was determined by measuring the *RiTEF* transcript abundance [[41]].

In Situ Hybridization with Digoxygenin (DIG) -Labeled LNA Probes

The *In situ* hybridization was carried out as described previously [[30]]. Before embedding the roots and nodules into paraffin they were fixed under vacuum using fixative (4% paraformaldehyde in PBS, pH7). LNA probes against mtr-miR171h were custom designed by Exiqon.

GUS Staining and Root Sectioning

β-glucuronidase (GUS) activity was used to assay miR171h and *NSP2* promoter activity as described in [[58],[59]]. Plants were grown under different conditions, as described earlier and harvest after 4 wpi. Shoots were removed and roots were incubated in GUS staining buffer containing 100 mM $NaPO_4$1 mM $K_3[Fe(CN)_6]$, 1 mM $K_4[Fe(CN)_6]$, 10 mM EDTA and 5-bromo-4-chloro-3-indolyl glucuronide, pH 7 at 37°C for different time scales, as indicated. Roots were embedded in 4% [w/v] agarose and 40 µm thick cross and longitudinal sections were obtained using the Leica VT 1000S vibratome (Bensheim, Germany).

Microscopic Analysis

Images of sections were generated using an Olympus BX41 microscope with the cellP software (Olympus, Hamburg, Germany) using different zoom objectives as indicated.

Images of tobacco leafs were performed using the Stereo-Fluorescence Microscope MZ 16FA (Leica) with a PLANATO 1×

objective (DC-DFC 300FX 1.3 MP, M.115×). Fluorescence proteins were excited with a mercury metal halide bulb laser, filtered with GFP3 filter system (exciting: 470/40 nm, barrier: 525/50 nm) for GFP fluorescence and a DsRED filter system (exciting: 545/30 nm, barrier: 620/60 nm) for mRFP fluorescence. Bright-field images were taken with identical settings except no laser-beam or filters were used. Exposure time for fluorescence pictures were 9 seconds with gain 5 and zoom factor 11.4×. Filters were successively shifted without changing settings. Image data were processed with the Leica LAS-AF Version 2.8.1.

Determination of mycorrhizal Parameters

Mycorrhizal colonization parameters of *Medicago* roots were determined by stained with Alexa Fluor® 488 Dye according to [[60]]. Roots were cleared with 10% [w/v] KOH for 5 min at 90°C. Roots were incubated at room temperature for at least 24 h in Alexa Fluor staining solution, containing 137 mM NaCl, 2.7 mM KCl, 12 mM Na_2HPO_4/ KH_2PO_4 and 0.01% [w/v] Alexa Fluor® 488 conjugate of wheat germ agglutinin (Invitrogen), and then washed five times prior mounting on microscopy slides. Mycorrhizal parameters defined by [[42]] are F%: Mycorrhizal frequency, M%: Mycorrhizal intensity per root, m%: Mycorrhizal intensity per root fragment A%: Arbuscule intensity per root, a% Arbuscule intensity per root fragment.

Protein Extraction and Western Blotting

Protein extraction and western blotting was carried out as described previously [[61]]. Protein concentration was determined using Bradford kit (Biorad) with provided γ-globulin standard following the manufacture's protocol.

Proteins were diluted to a final concentration of 1.5 μg/μl with homogenization buffer (100 mM HEPES, 10% [v/v] Glycerol, 5 mM dithiothritol, 1 tablet cOmplete™ ULTRA Tablets (Roche) per 10 ml buffer pH 7.5). 10 μl of protein solution was mixed with 10 μl of 2 × SDS sample buffer (125 mM Tris, 20% [v/v] glycerol, 4% [w/v] SDS, 0.01% [w/v] bromophenol blue, 40 μM DTE) and incubated for 5 min at 95°C. Samples were centrifuged at 14000 g for 1 min (4°C) and then resolved on a 1 mm 12% SDS polyacrylamide gel. The semi-dry blot was assembled as described in the Immobilon®-P Transfer Membrane (Millipore™) user guide. The blot was disassembled and the membrane was further used for immunostaining. Detection of GFP and mRFP was carried out using 1:2500 and 1:1000 dilutions for anti-GFP (rabbit, GenScript) and anti-mRFP (rabbit, Abcam®), respectively. As a loading control, a 1:10,000 diluted anti RubisCO antibody (rabbit, Agrisera™) was added to each of the antibody solutions. The secondary anti-rabbit-alkaline-phosphatase (Abcam®) antibody was diluted 1:5000. For visualization a ready-to-use NBT/BCIP solution (Roche®) was used. The membrane was dried and scanned with the Epson scanner Perfection 4870 (Epson Scan 3.0 Software). RubisCO, mRFP and GFP were identified by comparison to a pre-stained protein size marker (Thermo Scientific).

AUTHORS' CONTRIBUTIONS

VH, EAD, NG and AM performed biological experiments. EAD and FK designed and initiated the research. All authors analyzed the data. EAD and FK wrote the manuscript. All authors read and approved the final manuscript.

ACKNOWLEDGEMENTS

The work was supported by the Max-Planck-Society. We are grateful to Florian Frugier in providing us seeds of *cre1-1* and the respective WT controls as well as helpful comments. We would like to thank Ursula Krause for providing the anti-Rubisco antibody. Additionally, we would like to thank Daniela Zöller and Derek Nedveck for additional help in the sample preparations and Jean-Phillipe Combier, Nathan Pumplin as well as the anonymous reviewers for critical reading of the manuscript and their helpful comments.

REFERENCES

1. Harrison MJ: Molecular and cellular aspects of the arbuscular mycorrhizal symbiosis. *Annu Rev Plant Physiol Plant Mol Biol* 1999, 50:361-389.

2. Bucher M: Functional biology of plant phosphate uptake at root and mycorrhiza interfaces. *New Phytol* 2007, 173:11-26.

3. Javot H, Penmetsa RV, Terzaghi N, Cook DR, Harrison MJ: A *Medicago truncatula*phosphate transporter indispensable for the arbuscular mycorrhizal symbiosis. *Proc Natl Acad Sci U S A* 2007, 104:1720-1725.

4. Pumplin N, Harrison MJ: Live-cell imaging reveals periarbuscular membrane domains and organelle location in *Medicago truncatula* roots during arbuscular mycorrhizal symbiosis. *Plant Physiol* 2009, 151:809-819.

5. Oldroyd GE: Speak, friend, and enter: signalling systems that promote beneficial symbiotic associations in plants. *Nat Rev Microbiol* 2013, 11:252-263.

6. Akiyama K, Matsuzaki K, Hayashi H: Plant sesquiterpenes induce hyphal branching in arbuscular mycorrhizal fungi. *Nature* 2005, 435:824-827.

7. Maillet F, Poinsot V, Andre O, Puech-Pages V, Haouy A, Gueunier M, Cromer L, Giraudet D, Formey D, Niebel A, Martinez EA, Driguez H, Becard G, Denarie J: Fungal lipochitooligosaccharide symbiotic signals in arbuscular mycorrhiza. *Nature* 2011, 469:58-63.

8. Lin WY, Lin SI, Chiou TJ: Molecular regulators of phosphate homeostasis in plants. *J Exp Bot* 2009, 60:1427-1438.

9. Lopez-Raez JA, Charnikhova T, Gomez-Roldan V, Matusova R, Kohlen W, De Vos R, Verstappen F, Puech-Pages V, Becard G, Mulder P, Bouwmeester H: Tomato strigolactones are derived from carotenoids and their biosynthesis is promoted by phosphate starvation. *New Phytol* 2008, 178:863-874.

10. Yoneyama K, Yoneyama K, Takeuchi Y, Sekimoto H: Phosphorus deficiency in red clover promotes exudation of orobanchol, the signal for mycorrhizal symbionts and germination stimulant for root parasites. *Planta* 2007, 225:1031-1038.

11. Umehara M, Hanada A, Yoshida S, Akiyama K, Arite T, Takeda-Kamiya N, Magome H, Kamiya Y, Shirasu K, Yoneyama K, Kyozuka J, Yamaguchi S: Inhibition of shoot branching by new terpenoid plant hormones. *Nature* 2008, 455:195-200.

12. Liu W, Kohlen W, Lillo A, Op den Camp R, Ivanov S, Hartog M, Limpens E, Jamil M, Smaczniak C, Kaufmann K, Yang WC, Hooiveld GJ, Charnikhova T, Bouwmeester HJ, Bisseling T, Geurts R: Strigolactone biosynthesis in *Medicago truncatula* and rice requires the symbiotic GRAS-type transcription factors *NSP1* and *NSP2*. *Plant Cell* 2011, 23:3853-3865.
13. Kalo P, Gleason C, Edwards A, Marsh J, Mitra RM, Hirsch S, Jakab J, Sims S, Long SR, Rogers J, Kiss GB, Downie JA, Oldroyd GE: Nodulation signaling in legumes requires *NSP2*, a member of the GRAS family of transcriptional regulators. *Science* 2005, 308:1786-1789
14. Smit P, Raedts J, Portyanko V, Debelle F, Gough C, Bisseling T, Geurts R: *NSP1* of the GRAS protein family is essential for rhizobial Nod factor-induced transcription. *Science* 2005, 308:1789-1791.
15. Gobbato E, Marsh JF, Vernie T, Wang E, Maillet F, Kim J, Miller JB, Sun J, Bano SA, Ratet P, Mysore KS, Denarie J, Schultze M, Oldroyd GE: A GRAS-type transcription factor with a specific function in mycorrhizal signaling. *Curr Biol* 2012, 22:2236-2241.
16. Wang E, Schornack S, Marsh JF, Gobbato E, Schwessinger B, Eastmond P, Schultze M, Kamoun S, Oldroyd GE: A common signaling process that promotes mycorrhizal and oomycete colonization of plants. *Curr Biol* 2012, 22:2242-2246.
17. Oldroyd GE, Harrison MJ, Paszkowski U: Reprogramming plant cells for endosymbiosis. *Science* 2009, 324:753-754.
18. Voinnet O: Origin, biogenesis, and activity of plant microRNAs. *Cell* 2009, 136:669-687
19. Brodersen P, Sakvarelidze-Achard L, Bruun-Rasmussen M, Dunoyer P, Yamamoto YY, Sieburth L, Voinnet O: Widespread translational inhibition by plant miRNAs and siRNAs. *Science* 2008, 320:1185-1190.
20. Chen X: Small RNAs in development - insights from plants. *Cur Opin Genet Dev* 2012, 22:361-367.
21. Sunkar R, Li YF, Jagadeeswaran G: Functions of microRNAs in plant stress responses. *Trends Plant Sci* 2012, 17:196-203.

22. Guleria P, Mahajan M, Bhardwaj J, Yadav SK: Plant small RNAs: biogenesis, mode of action and their roles in abiotic stresses. *Dev Reprod Biol* 2011, 9:183-199.

23. Naqvi AR, Sarwat M, Hasan S, Roychodhury N: Biogenesis, functions and fate of plant microRNAs. *J Cell Physiol* 2012, 227:3163-3168.

24. Sun G: MicroRNAs and their diverse functions in plants. *Plant Mol Biol* 2012, 80:17-36.

25. Combier JP, Frugier F, de Billy F, Boualem A, El-Yahyaoui F, Moreau S, Vernié T, Ott T, Gamas P, Crespi M, Niebel A: *MtHAP2-1* is a key transcriptional regulator of symbiotic nodule development regulated by microRNA169 in *Medicago truncatula*. *Genes Dev* 2006, 20:3084-3088.

26. Boualem A, Laporte P, Jovanovic M, Laffont C, Plet J, Combier JP, Niebel A, Crespi M, Frugier F: MicroRNA166 controls root and nodule development in *Medicago truncatula*. *Plant J* 2008, 54:876-887.

27. Bazin J, Bustos-Sanmamed P, Hartmann C, Lelandais-Briere C, Crespi M: Complexity of miRNA-dependent regulation in root symbiosis. *Philos Trans R Soc Lond B Biol Sci* 2012, 367:1570-1579.

28. Branscheid A, Sieh D, Pant BD, May P, Devers EA, Elkrog A, Schauser L, Scheible WR, Krajinski F: Expression pattern suggests a role of MiR399 in the regulation of the cellular response to local Pi increase during arbuscular mycorrhizal symbiosis. *Mol Plant Microbe Interact* 2010, 23:915-926.

29. Gu M, Xu K, Chen A, Zhu Y, Tang G, Xu G: Expression analysis suggests potential roles of microRNAs for phosphate and arbuscular mycorrhizal signaling in *Solanum lycopersicum*. *Physiol Plant* 2010, 138:226-237.

30. Devers EA, Branscheid A, May P, Krajinski F: Stars and symbiosis: microRNA- and microRNA*-mediated transcript cleavage involved in arbuscular mycorrhizal symbiosis. *Plant Physiol* 2011, 156:1990-2010.

31. Branscheid A, Devers EA, May P, Krajinski F: Distribution pattern of small RNA and degradome reads provides information on miRNA gene structure and regulation. *Plant Signal Behav* 2011, 6:1609-1611.

32. Lauressergues D, Delaux PM, Formey D, Lelandais-Briere C, Fort S, Cottaz S, Becard G, Niebel A, Roux C, Combier JP: The microRNA miR171h modulates arbuscular mycorrhizal colonization of *Medicago truncatula* by targeting *NSP2*. *Plant J* 2012, 72:512-522.
33. Ariel F, Brault-Hernandez M, Laffont C, Huault E, Brault M, Plet J, Moison M, Blanchet S, Ichante JL, Chabaud M, Carrere S, Crespi M, Chan RL, Frugier F: Two direct targets of cytokinin signaling regulate symbiotic nodulation in *Medicago truncatula*. *Plant Cell* 2012, 24:3838-3852.
34. Gonzalez-Rizzo S, Crespi M, Frugier F: The *Medicago truncatula CRE1* cytokinin receptor regulates lateral root development and early symbiotic interaction with*Sinorhizobium meliloti*. *Plant Cell* 2006, 18:2680-2693.
35. Barker SJ, Tagu D: The roles of auxins and cytokinins in mycorrhizal symbioses. *J Plant Growth Regul* 2000, 19:144-154.
36. De Luis A, Markmann K, Cognat V, Holt DB, Charpentier M, Parniske M, Stougaard J, Voinnet O: Two microRNAs linked to nodule infection and nitrogen-fixing ability in the legume *Lotus japonicus*. *Plant Physiol* 2012, 160:2137-2154
37. Karimi M, Inze D, Depicker A: GATEWAY vectors for Agrobacterium-mediated plant transformation. *Trends Plant Sci* 2002, 7:193-195.
38. Nakagawa T, Kurose T, Hino T, Tanaka K, Kawamukai M, Niwa Y, Toyooka K, Matsuoka K, Jinbo T, Kimura T: Development of series of gateway binary vectors, pGWBs, for realizing efficient construction of fusion genes for plant transformation. *J Biosci Bioeng* 2007, 104:34-41.
39. Harrison MJ, Dewbre GR, Liu J: A phosphate transporter from *Medicago truncatula*involved in the acquisition of phosphate released by arbuscular mycorrhizal fungi. *Plant Cell* 2002, 14:2413-2429.
40. Breuillin F, Schramm J, Hajirezaei M, Ahkami A, Favre P, Druege U, Hause B, Bucher M, Kretzschmar T, Bossolini E, Kuhlemeier C, Martinoia E, Franken P, Scholz U, Reinhardt D:Phosphate systemically inhibits development of arbuscular mycorrhiza in *Petunia hybrida* and represses genes involved in mycorrhizal functioning. *Plant J* 2010, 64:1002-1017.

41. Helber N, Wippel K, Sauer N, Schaarschmidt S, Hause B, Requena N: A versatile monosaccharide transporter that operates in the arbuscular mycorrhizal fungus*Glomus sp* is crucial for the symbiotic relationship with plants. *Plant Cell* 2011, 23:3812-3823.

42. Trouvelot A, Kough JL, Gianinazzi-Pearson V: Mesure du taux de mycorhization VA d'un système radiculaire. Recherche des méthodes d'estimation ayant une signification fonctionnelle. In *The Mycorrhizae: Physiology and Genetic*. Edited by Gianinazzi-Pearson V, Gianinazzi S. INRA Presse, Paris; 1986:217-221.

43. Heckmann AB, Lombardo F, Miwa H, Perry JA, Bunnewell S, Parniske M, Wang TL, Downie JA:*Lotus japonicus* nodulation requires two GRAS domain regulators, one of which is functionally conserved in a non-legume. *Plant Physiol* 2006, 142:1739-1750.

44. Murakami Y, Miwa H, Imaizumi-Anraku H, Kouchi H, Downie JA, Kawaguchi M, Kawasaki S:Positional cloning identifies Lotus japonicus *NSP2,* a putative transcription factor of the GRAS family, required for *NIN* and *ENOD40* gene expression in nodule initiation. *DNA Res* 2006, 13:255-265.

45. Pant BD, Musialak-Lange M, Nuc P, May P, Buhtz A, Kehr J, Walther D, Scheible WR:Identification of nutrient-responsive Arabidopsis and rapeseed microRNAs by comprehensive real-time polymerase chain reaction profiling and small RNA sequencing. *Plant Physiol* 2009, 150:1541-1555.

46. Plet J, Wasson A, Ariel F, Le Signor C, Baker D, Mathesius U, Crespi M, Frugier F: *MtCRE1*-dependent cytokinin signaling integrates bacterial and plant cues to coordinate symbiotic nodule organogenesis in *Medicago truncatula*. *J* 2011, 65:622-633.

47. Salama AMSEA, Wareing PF: Effects of mineral-nutrition on endogenous cytokinins in plants of sunflower (*Helianthus-Annuus L*). *J Exp Bot* 1979, 30:971-981.

48. Kuiper D, Schuit J, Kuiper PJC: Effects of internal and external cytokinin concentrations on root-growth and shoot to root ratio of *Plantago-Major Ssp Pleiosperma* at different nutrient conditions. *Plant Soil* 1988, 111:231-236.

49. Gomez SK, Javot H, Deewatthanawong P, Torres-Jerez I, Tang Y, Blancaflor EB, Udvardi MK, Harrison MJ: *Medicago truncatula*

and *Glomus intraradices* gene expression in cortical cells harboring arbuscules in the arbuscular mycorrhizal symbiosis. *BMC Plant Biol* 2009, 9:10.

50. Hogekamp C, Arndt D, Pereira PA, Becker JD, Hohnjec N, Kuster H: Laser microdissection unravels cell-type-specific transcription in arbuscular mycorrhizal roots, including CAAT-box transcription factor gene expression correlating with fungal contact and spread. *Plant Physiol* 2011, 157:2023-2043.
51. Kawashima CG, Yoshimoto N, Maruyama-Nakashita A, Tsuchiya YN, Saito K, Takahashi H, Dalmay T: Sulphur starvation induces the expression of microRNA-395 and one of its target genes but in different cell types. *Plant J* 2009, 57:313-321
52. Pumplin N, Zhang X, Noar RD, Harrison MJ: Polar localization of a symbiosis-specific phosphate transporter is mediated by a transient reorientation of secretion. *Proc Natl Acad Sci U S A* 2012, 109:E665-E672.
53. Horvath B, Yeun LH, Domonkos A, Halasz G, Gobbato E, Ayaydin F, Miro K, Hirsch S, Sun J, Tadege M, Ratet P, Mysore KS, Ane JM, Oldroyd GE, Kalo P: *Medicago truncatula* IPD3 is a member of the common symbiotic signaling pathway required for rhizobial and mycorrhizal symbioses. *Mol Plant Microbe Interact* 2011, 24:1345-1358.
54. Auriac MC, Timmers AC: Nodulation studies in the model legume *Medicago truncatula*: advantages of using the constitutive *EF1alpha* promoter and limitations in detecting fluorescent reporter proteins in nodule tissues. *Mol Plant Microbe Interact* 2007, 20:1040-1047.
55. Nakagawa T, Suzuki T, Murata S, Nakamura S, Hino T, Maeo K, Tabata R, Kawai T, Tanaka K, Niwa Y, Watanabe Y, Nakamura K, Kimura T, Ishiguro S: Improved gateway binary vectors: High-performance vectors for creation of fusion constructs in Transgenic analysis of plants. *Biosci Biotech Bioch* 2007, 71:2095-2100.
56. Boisson-Dernier A, Chabaud M, Garcia F, Becard G, Rosenberg C, Barker DG: *Agrobacterium rhizogenes*-transformed roots of *Medicago truncatula* for the study of nitrogen-fixing and endomycorrhizal symbiotic associations. *Mol Plant Microbe Interact* 2001, 14:695-700.

57. Vaucheret H: Promoter-dependent trans-inactivation in transgenic tobacco plants - kinetic aspects of gene silencing and gene reactivation. *Cr Acad Sci Iii-Vie* 1994, 317:310-323.

58. Jefferson RA, Bevan M, Kavanagh T: The use of the *Escherichia-Coli* beta-glucuronidase gene as a gene fusion marker for studies of gene-expression in higher-plants. *Biochem Soc T* 1987, 15:17-18.

59. Jefferson RA, Kavanagh TA, Bevan MW: GUS fusions: beta-glucuronidase as a sensitive and versatile gene fusion marker in higher plants. *Embo J* 1987, 6:3901-3907

60. Grace C, Stribley DP: A safer procedure for routine staining of vesicular-arbuscular mycorrhizal fungi. *Mycol Res* 1991, 95:1160-1162.

61. Devers EA, Teply J, Reinert A, Gaude N, Krajinski F: An endogenous artificial microRNA system for unraveling the function of root endosymbioses related genes in *Medicago truncatula*. *BMC Plant Biol* 2013, 13:82.

Chapter 5

The Blue Fluorescent Protein from Vibrio Vulnificus Ckm-1 is a Useful Reporter for Plant Research

Jin-Min Tu[1], Ming-Chung Chang[2], Lynn LH Huang[3], Ching-Dong Chang[4], Hao-Jen Huang[5], Ruey-Hua Lee[1], and Ching-Chun Chang[1,3]

[1]Institute of Tropical Plant Sciences, National Cheng Kung University, 1 University Rd, Tainan 701, Taiwan

[2]Department of Nutrition, Hung Kuang University, Taichung 433 Taiwan

[3]Institute of Biotechnology, National Cheng Kung University, Tainan 701, Taiwan

[4]Department of Veterinary Medicine, National Pingtung University of Science and Technology, Pingtung 912, Taiwan

[5]Department of Life Sciences, National Cheng Kung University, Tainan 701, Taiwan

ABSTRACT

Background

The mBFP is an improved variant of NADPH-dependent blue fluorescent protein that was originally identified from the non-bioluminescent pathogenic bacteria *Vibrio vulnificus* CKM-1. To explore the application of mBFP in plants, the mBFP gene expression was driven by one of the three promoters, namely, leaf-specific (*RbcS*), hypoxia-inducible (*Adh*) or auxin-inducible (*DR5*) promoters, in different plant tissues such as leaves, roots and flowers under diverse treatments. In addition, the expressed mBFP protein was targeted to five subcellular compartments such as cytosol, endoplasmic reticulum, apoplast, chloroplast and mitochondria, respectively, in plant cells.

Results

When the mBFP was transiently expressed in the tobacco leaves and floral tissues of moth orchid, the cytosol and apoplast exhibited brighter blue fluorescence than other compartments. The recombinant mBFP-mS$_1$C fusion protein exhibited enhanced fluorescence intensity that was correlated with more abundant RNA transcripts (1.8 fold) as compared with a control. In the root tips of horizontally grown transgenic Arabidopsis, mBFP could be induced as a reporter under hypoxia condition. Furthermore, the mBFP was localized to the expected subcellular compartments, except that dual targeting was found when the mBFP was fused with the mitochondria-targeting signal peptide. Additionally, the brightness of mBFP blue fluorescence was correlated with NADPH concentration.

Conclusion

The NADPH-dependent blue fluorescent protein could serve as a useful reporter in plants under aerobic or hypoxic condition. However, to avoid masking the mitochondrial targeting signal, fusing mBFP as a fusion tag in the C-terminal will be better when the mBFP is applied in mitochondria trafficking study. Furthermore, mBFP might have the potential to be further adopted as a NADPH biosensor in plant cells. Future codon optimization of mBFP for plants could significantly enhance its brightness and expand its potential applications.

BACKGROUND

Fluorescent proteins have been widely used as specific markers to study gene and protein behaviors at the cellular, tissue, and organismal levels. The green fluorescent protein (GFP) from*Aequorea victoria* is the best studied and most widely used, and many other spectral variants have been derived from modification made to the GFP (Ai et al. [2007]; Cubitt et al. [1995]; Shaner et al. [2005]). Previously, blue fluorescent proteins (BFPs) derived from GFP were engineered by modifying the chromophore structures through the substitution of Tyr^{66} with a histidine (Ai et al.[2007]; Heim and Tsien [1996]; Lippincott-Schwartz and Patterson [2003]). However, several drawbacks of BFP variants that might limit their application have been reported, including low solubility, poor photostability, slow maturation times, and low quantum yields (Heim and Tsien[1996]; Lippincott-Schwartz and Patterson [2003]). Recently, the BFPs variants have been greatly improved which make them applicable in the imaging of mammalian cells (Ai et al. [2007]; Mena et al. [2006]; Subach et al. [2008], [2011]). For instance, Azurite, a BFP variant which is significantly enhanced in photostability and brightness (Mena et al. [2006]). Subsequently, several BFP derivatives that are even brighter than azurite, such as mKalama1, EBFP1.5, and EBFP2, were obtained, but only EBFP2 has better photostability. However, mKalama1 has a brighter fluorescence than EBFP2 due to the high efficiency in folding and chromophore maturation (Ai et al.[2007]). Alternatively, mTagBFP, a red fluorescent protein (RFP) derivative carrying a tyrosine-based chromophore has substantially higher brightness, faster chromophore maturation, and higher pH stability than BFPs with a histidine in the chromophore (Subach et al. [2008]). Meanwhile, mTagBFP2 exhibits greater chemical stability and photostability and substantially higher brightness in live cells, while maintaining the same other beneficial properties as mTagBFP (Subach et al. [2011]). However, BFPs and other GFP variants are not easily applicable in real-time imaging under anaerobic conditions due to the formation of chromophore through the oxidation of amino acid residues (Shaner et al. [2005]).

The oxygen-independent blue fluorescent proteins belonging to the short-chain dehydrogenase (SDR) family that have been identified either from a bacterial pathogen (Su et al. [2001]) or more recently from metagenomic DNA (Hwang et al. [2012]) might provide a alternative source for imaging live cells under both aerobic and anaerobic conditions. Previously, the NADPH-dependent blue fluorescent protein (BFPvv) gene was originally identified from the non-bioluminescent pathogen *Vibrio vulnificus* CKM-1 (Su et al. [2001]). Subsequently, random mutagenesis and DNA shuffling approaches have been applied to greatly increase the intensity and duration of BFPvv blue fluorescence (Su et al. [2001]; Chang et al. [2004a]; Kao et al. [2011]). The mechanism of the blue light emission from BFPvv is based on effectively augmenting the intrinsic fluorescence of bound NADPH (Chang et al. [2004b]; Kao et al. [2011]). The crystal structure of the BFPvvD8-NADPH complex has revealed that the increased fluorescent intensity of BFPvv is related to the conformational change caused by a Gly^{176} mutation close to the NADPH binding site (Kao et al.[2011]).

Due to the potential interference from auto-emission of blue fluorescence through the cinnamic acid covalently bound with the cell wall and other minor phenolic compounds present in the plant cell wall (Buschmann et al. [2000]), the BFPs of GFP variants have been rarely applied in plant research. To our best knowledge, the only previous study in addressing the application of BFP, a Y66H type of GFP variants have shown limited utility in monitoring viral movement in plants because its brightness was significantly influenced by the cellular pH environments among plant species (Diveki et al. [2002]). In this study, we explored the possibilities for the application of NADPH-dependent BFPvv from *Vibrio vulnificus* CKM-1 in plants. By expressing the improved variant of BFPvv in different subcellular compartments and tissue types among plant species, our results demonstrated that the BFPvv variant is a useful reporter for plant research.

METHODS

Construction of Plant Nuclear Expression Plasmids

The mBFP gene which was genetically modified variant of BFPvv from *Vibrio vulnificus* CKM-1 was PCR amplified and cloned into five different plant expression vectors of ImpactVector series (Wageningen UR, Netherlands). Additionally, the mBFP-mS_1C fusion gene, in which the mBFP gene was fused with the 3′ end region (528 bp) of the avian reovirus (ARV) *S1* gene (Lu et al. [2011]), was cloned into ImpactVector. In addition, the *Adh* promoter was PCR amplified from *Arabidopsis thaliana* ecotype Columbia according to a previous study (McKendree and Ferl [1992]), and was used to replace the *RbcS* promoter in the mBFP expression cassettes. Furthermore, the auxin-inducible DR5 promoter was synthesized (MDbio, Taiwan) according to a previous report (Ulmasov et al. [1997]), and was also used to replace the *RbcS* promoter in the mBFP expression cassettes. Subsequently, the mBFP or mBFP-mS_1C fusion gene expression cassettes driven by three different promoters in the ImpactVector-based plasmids were pasted into the PacI/AscI site of the pBINPLUS vector (Figure 1). To make sure no mutations were introduced during above PCR cloning, the newly constructed plasmids were sequenced. The PCR primers used in the cloning was listed. The plant expression vectors were transformed intoAgrobacteria tumerfaciens* GV3101 by electroporation.

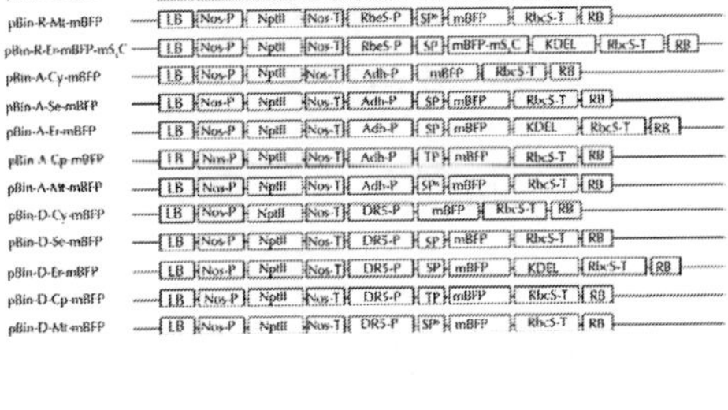

Figure 1: Schematic representation of plant expression vectors.The mBFP or mBFP-mS_1C fusion gene expression cassettes driven by three different pro-

moters in pBINPLUS vector were shown. Nos-P, *Nos*promoter; Nos-T, *Nos* terminator; NptII, Kanamycin resistance gene; RbcS-P, *RbcS* promoter; RbcS-T, *RbcS* terminator; Adh-P, *Adh* promoter; DR5-P, DR5 promoter; mBFP, variants of blue fluorescence protein (BFPvv) gene; mBFP-mS$_1$C, fusion of mBFP gene with 3′ end region of ARV *S1* gene; KDEL, ER retention signal; SP, signal peptide of secretory pathway; SP^m, mitochondrial signal peptide; TP, chloroplast transit peptide.

Agroinfiltration of Tobacco and Moth Orchid

The agroinfiltration procedure was performed according to a previous study (Sparkes et al.[2006]). The *Agrobacteria* carried with or without expression constructs were used to infiltrate tobacco (*Nicotiana tabacum* cv. Petit Havana) leaves and the petals of moth orchid (*Phalaenopsis aphrodite* subsp. *formosana*). The infiltrated plants were kept in dark for 2 days and then under light for 3 days except as indicated otherwise. The infiltrated tissues were treated with or without 50 μM NAA for 24 hr. Alternatively, the infiltrated floral tissues of moth orchid were treated with or without 1 mM H_2O_2 for 24 hr. The tissues were imaged under a fluorescence microscope (Olympus BX53, Japan) equipped with a UV light with an excitation wavelength of 330 ~ 385 nm and a filter with an emission wavelength of 420 nm. The images were captured by the Olympus U-TV1x-2 camera system under the control of SPOT imaging software (Spot Imaging Solutions, USA).

Transgenic Arabidopsis and Protoplast Preparation

The *Agrobacteria tumerfaciences* GV3101 carrying expression construct was used to transform*Arabidopsis thaliana* ecotype Columbia by floral dip method (Zhang et al. [2006]). The T_0transgenic plants were selected in MS media containing kanamycin (50 μg/ml) until T_2 or T_3generation as indicated. The integration of foreign DNA into nuclear genome of transgenic Arabidopsis was checked by PCR using KAPA3G plant PCR kits (Antec Bioscience, Taiwan) according to manufacturer's procedure. The seeds of transgenic Arabidopsis lines were plated onto the MS media containing kanamycin (50 μg/ml) and then grown horizontally or vertically in a growth chamber with 8/16 hr of dark/light cycle at

22°C. The protoplast preparation was referred to a previous report (Yoo et al. [2007]). In brief, the leaves of transgenic Arabidopsis or Agrobacteria-infiltrated tobacco leaves were sliced into about 0.5 mm segments and immersed into enzyme solution containing macerozyme R10 (Duchefa, Netherlands) and cellulose R10 (Yakult Honsha, Japan). After washing, the protoplasts were imaged under a confocal microscope (Zeiss LSM780).

RT-PCR Assay

Total RNA was isolated from infiltrated tobacco leaves by using a total RNA mini purification kit (BioKit, Taiwan) according to the manufacturer's procedure. After removing any trace amounts of DNA contamination, the RNA (1 μg) was reverse transcribed into cDNA with oligo-dT as the primer using a cDNA synthesis kit (Bioline, UK) according to the manufacturer's protocol. Subsequently, the gene-specific primers of mBFP and tobacco elongation factor 1α (*Nt*EF-1α) listed. The PCR condition was according to a previous study (Lu et al. [2011]), except 28 cycles was carried out. The PCR products were electrophoresed on a 0.8% agarose gel and then visualized by staining with ethidium bromide. The relative abundance of mBFP RT-PCR products to that of *Nt*EF-1α was further quantitated by Image J (National Institutes of Health, USA).

NADPH Treatment

The leaves of 4-week old wild type (WT) and transgenic Arabidopsis pBin-R-Se-mBFP line or pBin-R-Cy-mBFP line (Figure 1) were directly infiltrated with different concentrations of NADPH solution as indicated and kept at room temperature for 5 min before imaging. Alternatively, the protoplasts were isolated from infiltrated tobacco leaves, and incubated with 0, 0.25 or 0.5 μM NADPH in final concentration for 5 min, and subsequently the protoplasts were imaged using a confocal microscope.

RESULTS AND DISCUSSION

Construction of Plant Expression Vectors

To explore the potential application of BFPvv variants (mBFP) in plants, at least 16 plant expression vectors were constructed in which expression of mBFP or mBFP-mS_1C fusion constructs were driven by any one of the three promoters (*RbcS*, *Adh* and DR5) and the expressed proteins were targeted to five different subcellular compartments such as cytosol, apoplast, ER, chloroplast, or mitochondria, respectively (Figure 1). The mBFPvvD8 is an improved variant of BFPvv with mutagenesis in 11 amino acids (Kao et al. [2011]). More recently, mBFPvvD9 which can emit at least 30% brighter blue fluorescence than mBFPvvD8 in *E. coli*, was identified, and carried a point mutation at Arg^{199}. To compare the brightness of blue fluorescence between two different variants of mBFP in plant cells, the mBFPvvD8 and mBFPvvD9 were simultaneously and transiently expressed in opposite counterpart of the same tobacco leaves by agroinfiltration. In addition, the expressed mBFP variant proteins were targeted to cytosol (pBin-R-Cy-mBFP), apoplast (pBin-R-Se-mBFP) and mitochondria (pBin-R-Mt-mBFP), respectively (Figure 1). Our results showed that the mBFPvvD9 emitted much brighter blue light in all examined tissues than the mBFPvvD8 did (Figure 2A). Therefore, the mBFPvvD9 was further used in the following experiments.

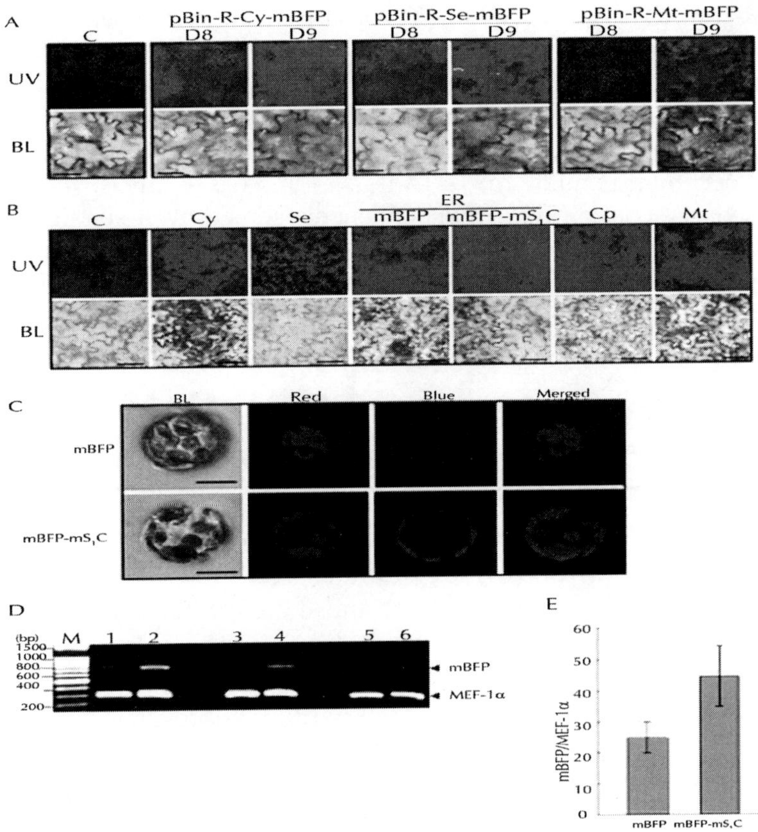

Figure 2: Transient expression of mBFP and mBFP-mS1C fusion protein in tobacco leaves by agroinfiltration. A. The mBFPvvD8 (D8) or mBFPvvD9 (D9) gene expression constructs driven by *RbcS*promoter were transiently expressed in tobacco leaves by agroinfiltration, respectively. The infiltrated tissues were imaged under fluorescence microscope with UV excitation (UV) or bright light (BL). C, *Agrobacteria*only as control. The magnification is 400 X and the ruler is 20 µm. B. The mBFP or mBFP-mS$_1$C fusion gene expression cassettes driven by *RbcS*promoter were transiently expressed in tobacco leaves by agroinfiltration, respectively, and the proteins were targeted to cytosol (Cy, pBin-R-Cy-mBFP), apoplast (Se, pBin-R-Se-mBFP), ER (mBFP, pBin-R-Er-mBFP; mBFP-mS$_1$C, pBin-R-Er-mBFP-mS$_1$C), chloroplast (Cp, pBin-R-Cp-mBFP) or mitochondria (Mt, pBin-R-Mt-mBFP), respectively. The infiltrated tissues were imaged under fluorescence microscope with UV excitation (UV) or bright light (BL). C, *Agrobacteria* only as control. The magnification is 200 X and the ruler is 40 µm. C. The protoplasts were isolated from tobacco leaves that transiently expressed pBin-R-Er-mBFP (mBFP) or

pBin-R-Er-mBFP-mS_1C (mBFP-mS_1C) constructs, and were imaged by confocal microscope. BL, bright light; Red, chlorophyll auto-fluorescence; Blue, mBFP blue fluorescence. The magnification is 1,000 X and the ruler is 20 μm. D. RT-PCR assay was used to measure the relative abundance of mBFP mRNA in tobacco leaves which transiently expressed pBin-R-Er-mBFP (Lanes 1, 3 and 5) or pBin-R-Er-mBFP-mS_1C (Lanes 2, 4 and 6) constructs. The *Nt*EF-1α was used as a control. Three independent experiments were carried out. M, 100 bp marker. E. The relative abundance of mBFP mRNA to that of *Nt*EF-1α was quantitated by Image J. mBFP, pBin-R-Er-mBFP construct; mBFP-mS_1C, pBin-R-Er-mBFP-mS_1C construct.

Transiently Expressed mBFP or Mbfp-Ms_1c Fusion Proteins in Tobacco Leaves

To study whether mBFP is a useful reporter in leaf tissues, the mBFP gene expression driven by*RbcS* promoter was transiently expressed in tobacco leaves by agroinfiltration, and the expressed mBFP proteins were targeted to five different subcellular compartments (Figure 1). The results showed that the expressed mBFP proteins no matter targeted to which compartments all can emit blue fluorescence under UV excitation as compared to a control (Figure 2B), but that stronger blue fluorescence was exhibited in the cytosol, apoplast and ER (Figure 2B). To test the effect of mBFP fusion protein on blue light emission, the mBFP was fused with the C-terminal region (176 amino acids) of ARV σC protein (Figure 1), which the C-terminal domain of σC protein was protease-stable and responsible for the cellular receptor binding (Calvo et al. [2005]). Then the ER-targeted mBFP or mBFP-mS_1C fusion protein was simultaneously and transiently expressed in the opposite counterpart of same tobacco leaves by agroinfiltration (Figure 2B). The results showed that mBFP-mS_1C fusion protein can emit much brighter blue fluorescence than mBFP protein can in leaf tissues (Figure 2B) or in protoplasts (Figure 2C). Since both expression constructs were driven by the same promoter, posttranscriptional regulation such as mRNA or protein stability could possibly be responsible for the difference. Therefore, RT-PCR assay was carried out to investigate the relative abundance of mRNA, and the results showed that plants transiently expressed mBFP-mS_1C fusion construct could accumulate about 1.8 fold more mRNA than that of non-fusion construct (Figure 2D and E). These results suggested that increasing mBFP mRNA stability could be a future strategy for increasing its expression level. Indeed, several potential cryptic introns and mRNA instability elements were present in the mBFP sequence; for instance, AGGAAGT, GTGGACGT, and GTGAACGT, the homologues of intron splicing donor sequences (AGGTAAGT or GTGTACGT), and ATTTG, the homologue of mRNA instability element (ATTTA), were found. In previous research, modifications of GFP by optimization of codon usage and elimination of a cryptic intron can greatly enhance the GFP expression in plants (Chiu et al. [1996]; Haseloff et al. [1997]). Therefore, to further improve the expression level of mBFP in plants, the codon usage might need to be optimized for plants in the future.

mBFP is a Useful Reporter Under Hypoxic Stress or Auxin Treatment

Further experiments were conducted to investigate whether or not the mBFP could be induced and imaged in different subcellular compartments and tissue types among plant species under various conditions. Firstly, five mBFP expression vectors driven by synthetic auxin-inducible DR5 promoter were constructed, and the expressed proteins were targeted to five different subcellular compartments (Figure 1). After agroinfiltration of tobacco leaves and flowers of moth orchid, they were then treated with or without 50 μM NAA for 24 hr. The results showed that mBFP blue fluorescence could be imaged in five subcellular compartments of tobacco leaf tissues and the petals of moth orchid, but that it showed brighter blue fluorescence in the cytosol, apoplast, and ER (Figure 3A and B). Secondly, five mBFP expression vectors driven by hypoxia-inducible *Adh*promoter were constructed, and the expressed proteins were targeted to five different subcellular compartments (Figure 1). Previously, it has been shown that H_2O_2 levels is significantly increased after O_2 deprivation and is required to trigger *Adh* gene expression (Baxter-Burrell et al. [2002]). Therefore, after agroinfiltration, the inflorescences of moth orchid were treated with or without 1 mM H_2O_2 for 24 hr. The results showed that although the mBFP blue fluorescence could be observed in all five subcellular compartments, the cytosol, apoplast, ER and chloroplasts showed brighter blue fluorescence (Figure 3C). Previously, it has been reported that the roots of horizontally rather than vertically grown

Arabidopsis will be under hypoxia pressure upon growing into the gel of a medium plate (Chung and Ferl [1999]). To investigate if mBFP was suitable for using as a reporter under hypoxic conditions, five transgenic Arabidopsis lines in which the mBFP expression was driven by *Adh* promoter, along with target proteins expected to accumulate in five different subcellular compartments, were generated. When the transgenic Arabidopsis lines were grown horizontally, the mBFP expression could be induced and the blue fluorescence could be clearly imaged in the root tips of transgenic pBin-A-Cy-mBFP (mBFP accumulated in cytosol) line, pBin-A-Se-mBFP (apoplast) line, pBin-A-Cp-mBFP (chloroplast) line and pBin-A-Mt-mBFP (mitochondria) line as compared with that in corresponding vertically grown plants (Figure 3D). These results suggested that mBFP is a useful reporter under hypoxic environments.

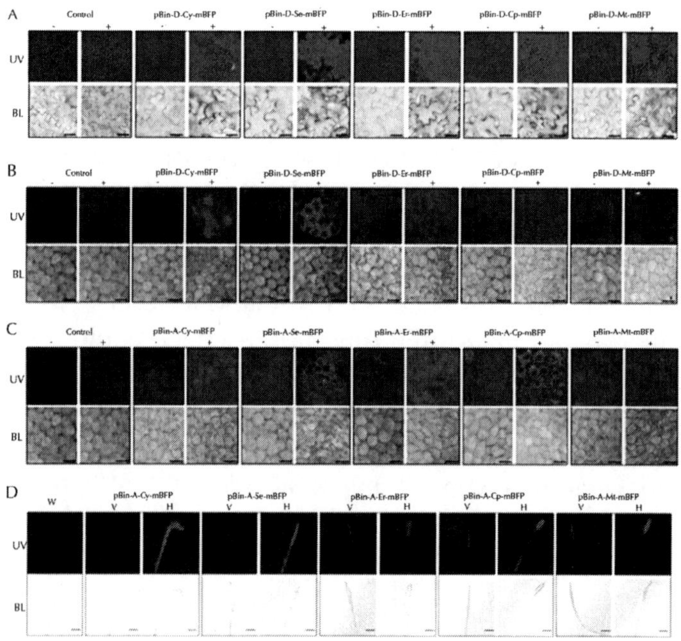

Figure 3: mBFP was induced as a reporter in different tissue types among plant species under various conditions. A. Tobacco leaves was infiltrated with Agrobacteria carrying mBFP expression constructs driven by DR5 promoter. The infiltrated tissues were further treated with or without 50 μM NAA for 24 hr. Subsequently, the tissues were imaged under fluorescence microscope

with UV excitation (UV) or bright light (BL). The induced mBFP proteins were targeted to cytosol (pBin-D-Cy-mBFP), apoplast (pBin-D-Se-mBFP), ER (pBin-D-Er-mBFP), chloroplast (pBin-D-Cp-mBFP) or mitochondria (pBin-D-Mt-mBFP), respectively. Control, *Agrobacteria* only; +, 50 μM NAA treatment; −, mock treatment. B. The flowers of moth orchid were carried out as the same as described in A. C. The mBFP expression cassettes driven by *Adh* promoter were transiently expressed in the petals of moth orchid. After agroinfiltration, the inflorescences were treated with or without 1 mM H_2O_2 for 24 hr. The induced mBFP proteins were targeted to cytosol (pBin-A-Cy-mBFP), apoplast (pBin-A-Se-mBFP), ER (pBin-A-Er-mBFP), chloroplast (pBin-A-Cp-mBFP) or mitochondria (pBin-A-Mt-mBFP), respectively. Control, *Agrobacteria* only; +, 1 mM H_2O_2 treatment; −, mock treatment. D. Five different transgenic Arabidopsis lines in which mBFP gene expression was driven by *Adh* promoter were generated by floral dip method, and the expressed proteins were targeted to cytosol (pBin-A-Cy-mBFP line), apoplast (pBin-A-Se-mBFP line), ER (pBin-A-Er-mBFP line), chloroplast (pBin-A-Cp-mBFP line) or mitochondria (pBin-A-Mt-mBFP line), respectively. The seeds of Arabidopsis were plated on MS media containing kanamycin, and subsequently were grown vertically (V) or horizontally (H) for 14 days after germination. The regions surrounding the root tips were imaged under fluorescence microscope with UV excitation (UV) or bright light (BL). W, wild type Arabidopsis. The magnification is 400 X (A), 200 X (B, C) or 40 X (D) and the ruler is 20 μm (A), 40 μm(B, C) or 1 mm (D).

Effect of NADPH on Intensity of mBFP Blue Fluorescence

NADPH is essential for mBFP blue fluorescence under UV excitation because the mechanism of mBFP blue fluorescence emission is due to the augmentation of the intrinsic bound NADPH fluorescence (Chang et al. [2004b]; Kao et al. [2011]). To investigate the effect of NADPH concentration on mBFP blue fluorescence *in planta*, leaf tissues of wild type or transgenic Arabidopsis pBin-R-Cy-mBFP and pBin-R-Se-mBFP lines in which the expressed mBFP protein was accumulated in cytosol and apoplast, respectively, were infiltrated with different NADPH concentrations (0, 0.1, 0.5, 1 and 2 μM) before imaging. The results showed that the higher the NADPH concentration, the stronger the mBFP blue fluorescence was emitted under UV excitation (Figure 4A). Alternatively, leaf disks of transgenic Arabidopsis (pBin-R-Se-mBFP line) were incubated with different concentrations of NADPH (0, 0.1,

0.5, 1 and 2 μM) for 15 min under suction before imaging. This result is consistent that mBFP blue fluorescence was significantly amplified with increasing NADPH concentrations. Furthermore, when the protoplasts isolated from tobacco leaves which mBFP was transiently expressed and targeted to cytosol were treated with different NADPH concentrations (0, 0.25 and 0.5 μM), brighter mBFP blue fluorescence was imaged in the presence of higher amounts of NADPH under confocal microscope analysis (Figure 4B). Although NADPH itself displays fluorescence which may interfere with the emission of fluorescence from mBFP-bounded NADPH, fluorescence from the free form of NADPH was not detectable under 0.5 μM NADPH (Figure 4B).

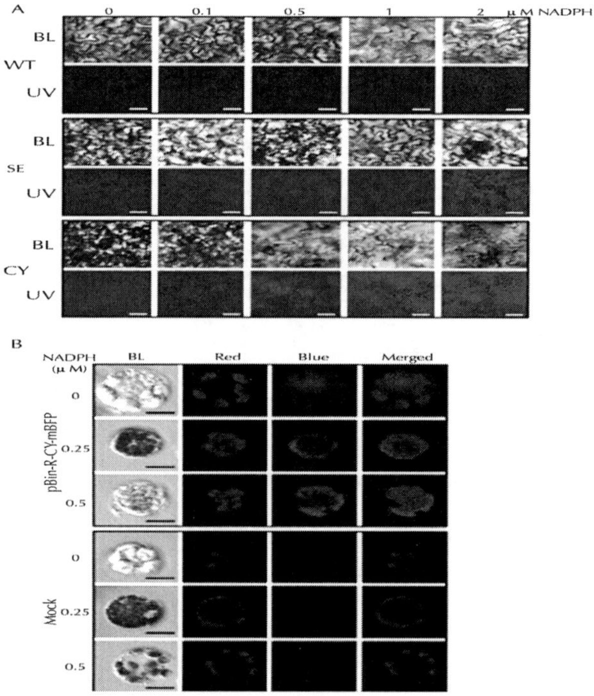

Figure 4: Effects of NADPH on the brightness of mBFP blue fluorescence. A. The leaves of wild type (WT) and transgenic Arabidopsis pBin-R-Se-mBFP (SE) line or pBin-A-Cy-mBFP (CY) line were directly infiltrated with NADPH as indicated concentration. Then the leaf tissues were imaged under fluorescence microscope with UV excitation (UV) or bright light (BL). The magnification is 400 X and the ruler is 20 μm. B. Tobacco leaves were infiltrated with Agrobacteria carrying pBin-R-Cy-mBFP construct or Agrobacteria only (Mock).

The protoplasts isolated from infiltrated leaves were treated with NADPH, and subsequently were imaged under confocal microscope. BL, bright light. Red, chlorophyll auto-fluorescence; Blue, mBFP blue fluorescence. The magnification is 1,000 X and the ruler is 20 μm.

Applications of the fluorescent proteins as biosensors are very useful for quantitative live imaging (Okumoto et al. [2012]). Previously, it has been shown that mBFPvvD8 could be applied in a Förster resonance energy transfer (FRET) assay (Kao et al. [2011]). Although the NADPH participates in many oxido-reductive reactions, and might be universally present in plant cells, its concentration may vary significantly in different subcellular compartments and tissue types. In this study, the brightness of mBFP fluorescence was shown to be positively correlated with the NADPH concentration (Figure 4). The results suggested that the mBFP might have the potential to be further developed as an NADPH biosensor *in planta*.

Subcellular Localization of mBFP by Confocal Microscope Analysis

Previous study have shown that signal peptide which helps to direct the target protein into subcellular compartments may be masked by the target protein itself, and thus causes mis-targeting or multiple targeting of the protein (Tanz et al. [2013]). In this study, mBFP accumulation in apoplast was obvious in the tested tissues (Figures 2, 3 and 4). To investigate if mBFP was correctly targeted into other compartments, the protoplasts were isolated from leaves of five transgenic Arabidopsis lines in which mBFP proteins were expected to accumulate in cytosol (pBin-R-Cy-mBFP line), ER (pBin-R-Er-mBFP and pBin-R-Er-mBFP-mS_1C lines), chloroplast (pBin-R-Cp-mBFP line) and mitochondria (pBin-R-Mt-mBFP line), respectively, and were imaged under confocal microscope. The ER-resident protein used the same secretory pathway as that of protein targeted to apoplast, except for having an additional ER-retention signal in the C-terminal. Although no ER-specific marker was used in this study, the mBFP or mBFP-mS_1C protein was seen to be intracellularly present, suggesting that it was ER-localized (Figures 2C and 5). The mBFP targeted to chloroplast was confirmed because of co-localization of mBFP blue fluorescence with chlorophyll auto-fluorescence (Figure 5). Therefore, our result showed that mBFP and mBFP-mS_1C fusion proteins were accumulated in the correct subcellular compartments except mitochondria (Figure 5), because the green fluorescence of MitoTracker was only partially co-localized with mBFP blue fluorescence in the protoplasts isolated from transgenic Arabidopsis (pBin-R-Mt-mBFP line) (Figure 5). Therefore, the protoplasts were isolated from tobacco leaves which transiently expressed pBin-R-Mt-mBFP construct for further investigation. The result was consistent with above that mBFP is partially co-localized with Mitotracker. It suggested that the mitochondria-targeted signal peptide might have been masked by mBFP, thus making it unable to be imported or inefficiently imported into mitochondria. Therefore, it will be better to avoid fusing mBFP in the N-terminal when apply a fusion mBFP tag protein for importing to plant mitochondria in future applications.

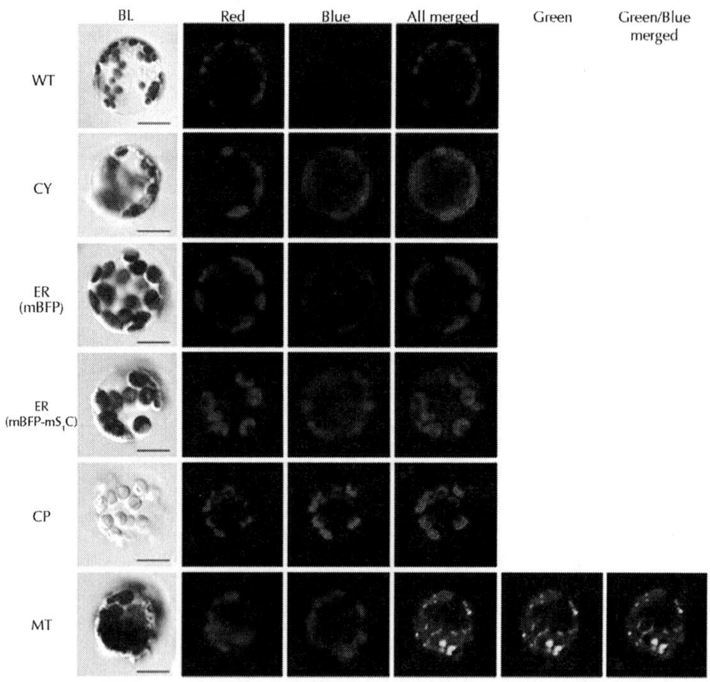

Figure 5: Subcellular localization of mBFP or mBFP-mS1C fusion protein in protoplasts of transgenic Arabidopsis. Five different transgenic Arabidopsis lines in which the expression of mBFP or mBFP-mS_1C gene was driven by *RbcS* promoter were generated by floral dip method, and the expressed mBFP or mBFP-mS_1C fusion proteins were expected to accumulate in the cytosol

(CY, pBin-R-Cy-mBFP), ER (mBFP, pBin-R-Er-mBFP; mBFP-mS$_1$C, pBin-R-Er-mBFP-mS$_1$C), chloroplast (CP, pBin-R-Cp-mBFP) or mitochondria (MT, pBin-R-Mt-mBFP), respectively. The protoplasts were isolated from the leaf tissues of transgenic Arabidopsis lines and wild type (WT), and then were imaged under confocal microscope. Alternatively, the protoplasts isolated from transgenic pBin-R-Mt-mBFP line were treated with MitoTracker (Invitrogen) before imaging. BL, bright light; Red, chlorophyll auto-fluorescence; Blue, mBFP blue fluorescence; Green, green fluorescence of MitoTracker. The magnification is 1,000 X and the ruler shown is 20 μm in length.

CONCLUSIONS

Through expression of mBFP or mBFP-mS$_1$C fusion protein in different subcellular compartments and tissue types among three different plant species under various conditions, in this study, we demonstrated that mBFP protein from *Vibrio vulnificus* CKM-1 is a useful reporter *in planta*. It could be used, for example, as a fusion tag to study protein trafficking and localization under both aerobic and anaerobic environments. In addition, the brightness of mBFP blue fluorescence was positively correlated to the NADPH concentration which might suggest that mBFP has great potential to be further developed as a NADPH biosensor for plant research. Although the blue auto-fluorescence coming from secondary cell wall might limit the utility of blue fluorescence protein in plant research, however, when the experiments were carried out with strict control, it will be applicable and useful as a reporter as shown in this study. Furthermore, further improvement of mBFP through codon optimization for plants might significantly enhance its brightness and expand its potential applications.

AUTHORS' CONTRIBUTIONS

The presented research work was carried out in collaboration between authors. LLHH, CDC, CCC designed research; JMT, MCC performed research work; JMT, HJH, RHL, CCC analyzed data; and CCC wrote the paper. All authors read and approved the final manuscript.

ACKNOWLEDGEMENTS

We appreciated Dr. Feng Teng-Yung for sharing the plasmids. We also thank Mr. Lee Yueh-Feng for help. We are also grateful to the two anonymous reviewers for their comments that help to improve the quality of this manuscript. This work was financially supported in part by the grant (MOST 103-2321-B-006-009) to C.-C. Chang from the Ministry of Science and Technology, Taiwan.

REFERENCES

1. Ai HW, Shaner NC, Cheng Z, Tsien RY, Campbell RE (2007) Exploration of new chromophore structures leads to the identification of improved blue fluorescent proteins. Biochem 46:5904-5910
2. Baxter-Burrell A, Yang Z, Springer PS, Bailey-Serres J (2002) RopGAP4-dependent Rop GTPase rheostat control of Arabidopsis oxygen deprivation tolerance. Science 296:2026-2028
3. Buschmann C, Langsdorf G, Lichtenthaler HK (2000) Imaging of the blue, green, and red fluorescence emission of plants: an overview. Photosynthetica 38:483-491
4. Calvo PG, Fox GC, Parrado XLH, Llamas-Saiz AL, Costas C, Martinez-Costas J, Benavente J, Raaij MJV (2005) Structure of the carboxy-terminal receptor-binding domain of avian reovirus fibre sigmaC. J Mol Biol 354:137-149
5. Chang MC, Chang CC, Chuang YC (2004) Fluorescent intensity of a novel NADPH-binding protein of *Vibrio vulnificus* can be improved by directed evolution. Biochem Biophys Res Commun 322:303-309
6. Chang YC, Chang CC, Chuang YC, Chen YC, Chang MC (2004) Bright fluorescence of a novel protein from *Vibrio vulnificus* depends on NADPH and the expression of this protein is regulated by a LysR-type regulatory gene. Biochem Biophys Res Commun 319:207-213
7. Chiu WL, Niwa Y, Zeng W, Hirano T, Kobayashi H, Sheen J (1996) Engineered GFP as a vital reporter in plants. Curr Biol 6:325-330

8. Chung HJ, Ferl RJ (1999) Arabidopsis alcohol dehydrogenase expression in both shoots and roots is conditioned by root growth environment. Plant Physiol 121:429-436
9. Cubitt AB, Heim R, Adams SR, Boyd AE, Gross LA, Tsien RY (1995) Understanding, improving and using green fluorescent proteins. Trends Biochem Sci 20:448-455
10. Diveki Z, Salanki K, Balazs E (2002) Limited utility of blue fluorescent protein (BFP) in monitoring plant virus movement. Biochimie 84:997-1002
11. Haseloff J, Siemering KR, Prasher DC, Hodg S (1997) Removal of a cryptic intron and subcellular localization of green fluorescent protein are required to mark transgenic Arabidopsis plants brightly. Proc Natl Acad Sci U S A 94:2122-2127
12. Heim R, Tsien RY (1996) Engineering green fluorescent protein for improved brightness, longer wavelengths and fluorescence resonance energy transfer. Curr Biol 6:178-182
13. Hwang CS, Choi ES, Han SS, Kim GJ (2012) Screening of a highly soluble and oxygen-independent blue fluorescent protein from metagenome. Biochem Biophys Res Commun 419:676-681
14. Kao TH, Chen Y, Pai CH, Chang MC, Wang HJ (2011) Structure of a NADPH-dependent blue fluorescent protein revealed the unique role of Gly176 on the fluorescence enhancement. J Struct Biol 174:485-493
15. Lippincott-Schwartz J, Patterson GH (2003) Development and use of fluorescent protein markers in living cells. Science 300:87-91
16. Lu SW, Wang KC, Hj L, Chang CD, Huang HJ, Chang CC (2011) Expression of avian reovirus minor capsid protein in plants. J Virol Methods 173:287-293
17. McKendree WL, Ferl RJ (1992) Functional elements of the Arabidopsis *Adh* promoter include the G-box. Plant Mol Biol 19:859-862
18. Mena M, Treynor TP, Mayo SL, Daugherty PS (2006) Blue fluorescent proteins with enhanced brightness and photostability from a structurally targeted library. Nat Biotechnol 24:1569-1571
19. Okumoto S, Jones A, Frommer WB (2012) Quantitative imaging with fluorescent biosensors. Annu Rev Plant Biol 63:663-706

20. Shaner NC, Steinbach PA, Tsien RY (2005) A guide to choosing fluorescent proteins. Nat Methods 2:905-909
21. Sparkes IA, Runions J, Kearns A, Hawes C (2006) Rapid, transient expression of fluorescent fusion proteins in tobacco plants and generation of stably transformed plants. Nat Protoc 1:2019-2025
22. Su JH, Chuang YC, Tsai YC, Chang MC (2001) Cloning and characterization of a blue fluorescent protein from *Vibrio vulnificus*. Biochem Biophys Res Commun 287:359-365
23. Subach OM, Gundorov IS, Yoshimura M, Subach FV, Zhang J, Gruenwald D, Souslova EA, Chudakov DM, Verkhusha VV (2008) Conversion of red fluorescent protein into a bright blue probe. Chem Biol 15:1116-1124
24. Subach OM, Cranfill PJ, Davidson MW, Verkhusha VV (2011) An enhanced monomeric blue fluorescent protein with the high chemical stability of the chromophore. PLoS One 6:e28674
25. Tanz SK, Castleden I, Small ID, Millar AH (2013) Fluorescent protein tagging as a tool to define the subcellular distribution of proteins in plants. Front Plant Sci 4:1-9
26. Ulmasov T, Murfett J, Hagen G, Guilfoyle TJ (1997) Aux/Iaa proteins repress expression of reporter genes containing natural and highly active synthetic auxin response elements. Plant Cell 9:1963-1971
27. Yoo SD, Cho YH, Sheen J (2007) Arabidopsis mesophyll protoplasts: a versatile cell system for transient gene expression analysis. Nat Protoc 2:1565-1572
28. Zhang X, Henriques R, Lin SS, Niu QW, Chua NH (2006) Agrobacterium-mediated transformation of *Arabidopsis thaliana* using the floral dip method. Nat Protoc 1:1-6.

Chapter 6

Shading During the Grain Filling Period Increases 2-Acetyl-1-Pyrroline Content in Fragrant Rice

Zhaowen Mo[1,2], Wu Li[3], Shenggang Pan[1,2], Timothy L Fitzgerald[4], Feng Xiao[1], Yongjian Tang[1], Yilei Wang[1], Meiyang Duan[1,2], Hua Tian[1,2], and Xiangru Tang[1,2]

[1]College of Agriculture, South China Agricultural University, Guangzhou 510642, Guangdong, China

[2]Scientific Observing and Experimental Station of Crop Cultivation in South China, Ministry of Agriculture. P. R. China, Guangzhou 510642, Guangdong, China

[3]Crops Research Institute, Guangdong Academy of Agricultural Sciences, Guangzhou 510640, Guangdong, China

[4]CSIRO Agriculture Flagship, Queensland Bioscience Precinct, 306 Carmody Rd, St Lucia 4067, Queensland, Australia

ABSTRACT

Background

Fragrant rice, including Thai jasmine and Indian basmati varieties, is highly valued by consumers globally. 2-acetyl-1-proline (2-AP) is the major compound responsible for the aromatic character of fragrant rice. Previously, environmental factors such as water management and salinity have been proven to influence 2-AP levels in fragrant rice; assessing the effect of additional environmental factors on 2-AP concentration is therefore eminent. The level of solar radiation (solar intensity; SI) to which a crop is exposed can affect growth, yield and grain quality, and other photosynthetic and physiological characteristics. In this study the effect of shading (i.e. the reduction of SI) on yield, quality, and 2-AP concentration in two elite Chinese fragrant rice varieties, 'Yuxiangyouzhan' and 'Nongxiang 18', has been investigated. Furthermore, accumulation of the plant stress response molecules proline and gamma-aminobutyric acid, which have also been implicated in pathways leading to 2-AP production, was assessed to study shading effects on these compounds in fragrant rice, and to further possibly determine fluxes in biochemical pathways leading to 2-AP accumulation.

Results

This study has revealed significant changes in the yield and quality characters under shading treatment. Additionally, 2-AP and GABA content in grains was significantly increased for all shading treatments in both varieties. In addition to 2-AP, ten other volatile compounds were studied; results indicated that shading treatments could have a selective effect on the metabolism of these volatile compounds.

Conclusions

In this study, we have demonstrated that shading during grain filling has significant effects on yield and quality traits in rice, and leads to the accumulation of GABA and 2-AP. We discuss the implications of these findings in terms of pathways leading to 2-AP and GABA production in fragrant rice, which have not been fully elucidated. The shading effect on ten additional volatile compounds is also discussed. Finally we discuss possible effects of variation in solar intensity resulting from anthropogenic emissions on fragrant rice production.

BACKGROUND

Fragrant rice varieties possess a characteristic aroma that is often described as 'nutty' or 'popcorn-like' (Bryant and McClung, [2010]). Thai 'jasmine' and indian 'basmati' are the most widely-recognized types of fragrant rice. Fragrant rice is highly desired by consumers and attracts premium price in many markets (Sakthivel et al., [2009]); furthermore, global demand for fragrant rice is increasing (Hashemi et al., [2013]).

The volatile compounds in rice are complicated; instrumental analyses have observed more than 200 volatile compounds (Tsugita, [1985]; Buttery et al., [1988]; Champagne, [2008]). Buttery et al. ([1988]) demonstrated that the probable key contributors to cooked rice aroma among the detected compounds were 2-acetyl-l-pyrroline, (E, E)-2, 4-decadienal, nonanal, hexanal, (E)-2-nonenal, octanal, decanal, 4-vinyl-guaiacol, and 4-vinylphenol. Jezussek et al. ([2002]) suggested that 2-amino acetophenone and 4, 5-epoxy-(E)-2-decenal was important previously unknown rice aroma compounds. Maraval et al. ([2008]) indicated that organic extracts of cooked aromatic rice showed similar aroma profile between the cultivars, but showed substantial differences in the levels of various volatile compounds. Moreover, Yang et al. ([2008]) suggested that 13 volatile coupounds may contribute to differences in aroma.

Despite the complexity of volatile compounds in rice (Champagne et al., [2005]), a single volatile compound, 2-acetyl-1-pyrroline (2-AP), is well-proven to be primarily responsible for the aromatic character of fragrant rice (Buttery et al., [1983], [1986]; Widjaja et al., [1996]; Jezussek et al.,[2002]). In fragrant rice, 2-AP can be detected in all parts of the plant except the roots (Buttery et al., [1983]; Maraval et al., [2010]); 2-AP is also present in non-fragrant rice varieties, but at much lowers concentration (Widjaja et al., [1996]).

The inactivation of *BADH2*, encoding an aminoaldehyde dehydrogenase, is responsible for the accumulation of 2-AP in fragrant rice varieties (Bradbury et al., [2005]; Bradbury et al., [2008]). Proline is a precursor of 2-AP in rice (Yoshihashi et al., [2002]), and γ-aminobutyraldehyde (GABald) is likely to be the direct precursor to 2-AP production via *BADH2*; Bradbury et al. ([2008]) proposed that functional *BADH2* catalyses the conversion of GABald to γ-aminobutyric acid (GABA), while non-functional *BADH2* in fragrant rice leads to cyclisation of GABald to Δ^1-pyrroline, with the addition of an acetyl group forming 2-AP. The role of proline in plant stress tolerance is well-established (Szabados and Savouré, [2010]); additionally, GABA accumulates in response to biotic and abiotic stresses in plants (Bouche and Fromm, [2004]) and is therefore speculated to play a role in stress tolerance (Kinnersley and Turano, [2000]). Thus, it appears that there is an overlap in pathways involved in 2-AP accumulation and the response to stress. In further support of an interaction between 2-AP production and stress responses, environmental factors have been observed to influence the level of 2-AP in fragrant rice grains. For example, Yoshihashi et al. ([2002]) reported that drought stress during grain formation increased 2-AP content, and Gay et al. ([2010]) and Poonlaphdecha et al. ([2012]) have also recently reported an increase in 2-AP accumulation in response to salt stress. Interestingly, elite fragrant rice varieties are relatively susceptible to abiotic and biotic stresses (Niu et al., [2008]), and recently an association of the fragrance phenotype with salt susceptibility has also been reported (Fitzgerald et al., [2010]; Wijerathna et al., [2014]).

'Photosynthetically Active Radiation' (PAR) is the range of electromagnetic radiation between 400 and 700 nanometers, roughly equivalent to the visible light spectrum, that plants (and most photosynthetic organisms) harness for photosynthesis. The level of PAR to which a plant is exposed is related to photosynthetic rate (McCree, [1981]), and variation in PAR can therefore have a major effect on plant growth and development. In the field, the intensity of solar radiation (solar intensity; SI) determines the level of PAR to which a crop is exposed, and variation in solar radiation can affect crop yield (Liu and Tollenaar, [2009]). Season, geographic location, and cloud cover are key factors influencing SI; additionally, SI varies in response to atmospheric pollution and water vapour content (Jáuregui and Luyando, [1999]; Haywood et al., [2011]). The level of SI to which a plant is exposed can also be strongly affected by surrounding vegetation; 'shading stress' in plants growing beneath or amongst a canopy, and responses to such stress, are well-recognised phenomena (Gommers et al., [2013]).

To understand how rice responds to low light environment, researchers have used artificial shading to control light density. Studies have showed that shading can affect rice morphological characteristics, physiological characteristics, yield, and quality. For example, Tang ([1988]) found that shading affected growth and yield, decreased dry matter, and increased leaf thickness. Cai and Luo ([1999]) reported that shading at different stages decreased rice yield, accumulation rate of dry matter and the uptake of nutrients, but increased nutrient content. Ren et al. ([2003]) indicated that the distribution and transformation of plant dry matter is affected by shading, and some shading levels decreased rice yield. Besides, Deng et al. ([2009]) found that the filled grain percentage and yield were obviously affected by shading. Moreover, Liu et al. ([2012]) reported that the effect of shading on the content of malondialdehyde, souble sugar, souble protein, and protective enzymes activities in leaves various in genotypes. Ding et al. ([2004]) found that shading after heading extended rice growth duration of the flag leaf, decreased decomposing rate of chlorophyll and MDA content, reduced the photosynthesis of the flag leaf, and that the degree of the grain-filling was lower than that of under the strong light. Zhang et al. ([2007]) indicated that shading affected rice quality differently amongst genotypes, however shading increased protein content and chalkiness and decreased amylose content of all rice cultivars. Liu et al. ([2008]) found that weak light during grain filling affected rice yield, physiological characteristics and quality, particularly the rate of chalky grains. Additionally, shading has been demonstrated to effect starch synthase and related enzyme activities (Wang et al., [2013]).

Previous studies have assessed the effects of an artificial shading (which results in a decrease in SI and PAR) on growth and development, yield, and grain quality in rice (e.g. Tanaka and Kawano,[1966]; Cai and Luo, [1999]; Ren et al., [2003]; Zhang et al., [2007]). Here we have studied the impact of shading during the grain filling period on GABA, proline, and 2-AP accumulation in fragrant rice. Furthermore, to provide additional insight into the effect of shading on growth and yield in fragrant rice, we have complemented our analyses of metabolite accumulation with assessment of total nitrogen levels, and grain yield and quality in response to shading treatment.

RESULTS

Effect of Shading Treatment on Yield, Yield Related Traits, Dry Weight, and Harvest Index

For Yuxiangyouzhan, all shading treatments resulted in a significant reduction in filled grain percentage, 1000-grain weight, grain yield, total dry weight, and harvest index. Additionally, shading treatment during the whole grain filling period (S1) and during the latter stage of grain filling (S3) caused a significant reduction of panicle number. For Nongxiang 18, shading treatment during the whole grain filling stage S1 and during early grain filling (S2) caused a significant reduction of filled grain percentage and 1000-grain weight; and S1 significantly decreased grain yield and total dry weight (Table 1). On average, Yuxiangyouzhan had lower panicles number, filled grain percentage, 1000-grain weight, and total dry weight, but had a higher number of grains per panicle, grain yield, and harvest index than Nongxiang 18 (Table 1).

Table 1: Effect of shading treatment on yield, yield related traits, total dry weight, and harvest index

Treatment	Panicles number (m^2)	Grains per panicle	Filled grain percentage (%)	1000-grain weight (g)	**Grain yield ($g.m^{-2}$)**	**Total dry weight ($g.m^{-2}$)**	**Harvest index (%)**
Yuxiangy ouzhan							
S0	263.33 a	185.09 a	84.48 a	21.27 a	863.14 a	1611.0 a	53.613 a
S1	190.00 c	162.63 a	51.61 c	19.51 b	494.04 c	1021.0 c	48.311 b
S2	250.00 ab	165.47 a	63.38 b	19.58 b	660.66 b	1328.5 b	49.602 b
S3	218.33 bc	179.61 a	64.74 b	20.04 b	523.69 c	1100.4 c	47.512 b
mean	230.42	173.20*	66.05	20.10	635.38	1265.2	49.759*

Nongxiang 18							
S0	284.42 a	148.41 a	85.61 a	25.46 a	655.19 a	1453.8 a	44.783 a
S1	261.67 a	126.64 a	53.32 c	22.94 b	457.25 b	1100.4 b	41.495 a
S2	290.00 a	129.09 a	66.01 b	23.07 b	639.96 a	1394.9 a	45.841 a
S3	273.33 a	148.32 a	81.76 a	24.84 a	638.35 a	1376.3 a	46.084 a
mean	277.35*	138.11	71.68	24.08*	597.68	1331.4	44.551

Means in the same column followed by different lower case letters for the same variety differ significantly at $P=0.05$ by LSD tests. Means of the two varieties followed by asterisk for the same detected index difer significant at $P=0.05$ by LSD tests.

Mo *et al.*

Mo *et al. Rice*2015 8:9, doi: 10.1186/s12284-015-0040-y

Effect of Shading Treatment on Grain Quality

Significant effects were identified for some shading treatments on most quality traits. All shading treatments significantly increased grain protein content in both Yuxiangyouzhan and Nongxiang 18 (Table 2). Additionally, a similar effect of shading on chalkiness characterstics was observed in both cultivars; shading during the whole grain filling period and during late grain filling resulted in a decrease in mean rate of chalky grains, and mean degree of chalkiness, however shading during early grain filling significantly increased chalkiness characteristics (Table 2). No significant effect on brown rice rate was detected in the shading treatments for either cultivar, while for milled rice rate, head rice rate and alkali spreading value effects varied with timing, during shading as well as cultivar (Table 2).

Table 2: Effect of shading treatment on grain quality

Treat ment	Brown rice rate (%)	Milled rice rate (%)	Head rice rate (%)	**Protein content (%)**	**Amylose content (%)**	**Alkali**	**grains with chal kiness (%)**	**Chal kiness degree (%)**
Yuxiangyouzhan								
S0	83.81 a	73.00 ab	70.25 a	9.53 c	23.43 c	7.47 ab	23.33 b	8.63 b
S1	83.56 a	72.66 bc	69.99 a	10.23 b	23.67 c	7.37 b	14.33 c	7.33 b
S2	83.62 a	72.29 c	68.29 b	10.40 a	24.90 b	7.57 a	32.00 a	18.17 a
S3	83.71 a	73.14 a	70.48 a	9.23 d	25.97 a	7.40 ab	18.67 bc	7.70 b
mean	83.67*	72.77*	69.75*	9.85*	24.49*	7.45*	22.08*	10.46*
Nongxiang 18								
S0	81.97 a	68.40 b	64.43 b	8.43 d	19.20 a	6.60 a	4.33 b	1.30 b
S1	82.43 a	69.21 ab	64.29 b	9.17 a	18.23 a	6.60 a	2.33 c	0.71 c
S2	81.64 a	68.58 b	64.57 b	8.97 b	18.50 a	6.63 a	13.67 a	3.00 a
S3	82.52 a	69.77 a	66.39 a	8.60 c	18.50 a	6.53 a	2.33 c	0.73 c
mean	82.14	68.99	64.92	8.79	18.61	6.59	5.67	1.44

Means in the same column followed by different lower case letters for the same variety differ significantly at $P=0.05$ by LSD tests. Means of the two varieties followed by asterisk for the same grain quality trait difer significant at $P=0.05$ by LSD tests.

Mo *et al.*

Mo *et al.* *Rice* 2015 8:9, doi: 10.1186/s12284-015-0040-y

Effect of Shading Treatment on 2-Acetyl-1-Pyrroline, GABA, Proline, and Total Nitrogen Levels

The 2-AP content in grains was significantly increased for all shading treatments in both varieties, with increases of 59.07 – 106.65%, and 11.89 – 42.37% detected for Yuxiangyouzhan and Nongxiang 18, respectively. In both varieties, the highest mean 2-AP content in grains was identified for plants exposed to S2 treatment, with 2-AP contents of 175.86 $\mu g \cdot kg^{-1}$ in Yuxiangyouzhan and 135.02 $\mu g \cdot kg^{-1}$ in Nongxiang 18, compared to respective means of 85.10 and 94.84 $\mu g \cdot kg^{-1}$ in the absence of shading (S0) (Figure 1). All shading treatments also significantly increased GABA in the grains of Yuxiangyouzhan (23.59 - 31.01%) and Nongxiang 18 (56.99 - 94.18%) (Figure 2). In Yuxiangyouzhan, a significant increase in proline content was identified for S1 and S2, but not for S3 (Figure 3). For Nongxiang 18, proline content was not significantly affected by any shading treatment (Figure 3). No significant difference in total nitrogen content in grains between shading treatments was observed (Figure 4).

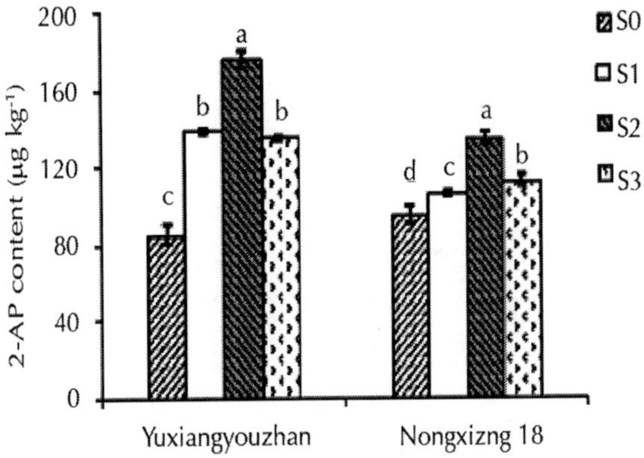

Figure 1: Effect of shading treatment on grain 2-AP content in grains. Vertical bars with different lower case letters above are significantly different at $P=0.05$ by LSD tests. Capped bars represent SD.

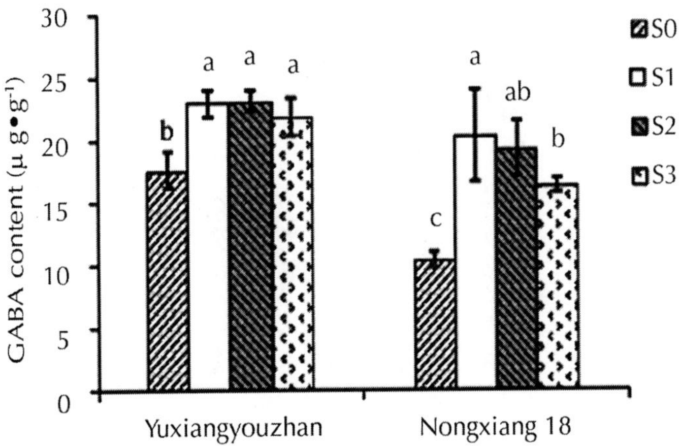

Figure 2: Effect of shading treatment on grain GABA content in grains. Vertical bars with different lower case letters above are significantly different at $P=0.05$ by LSD tests. Capped bars represent SD.

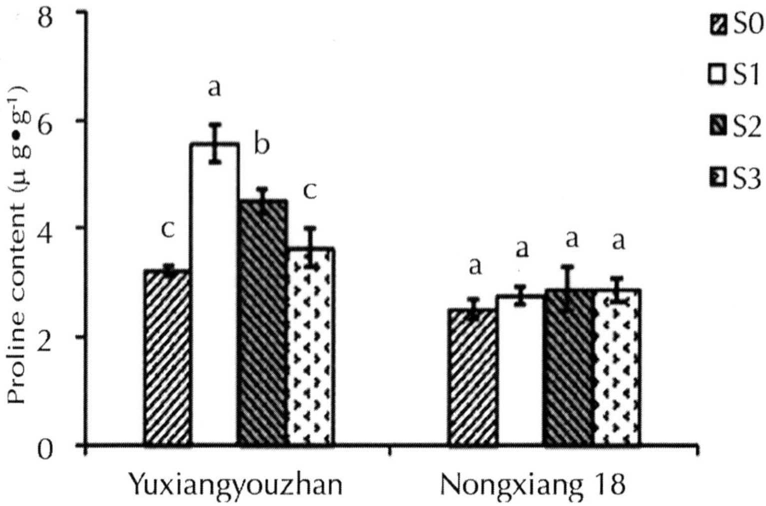

Figure 3: Effect of shading treatment on proline content in grains.Vertical bars with different lower case letters above are significantly different at $P=0.05$ by LSD tests. Capped bars represent SD.

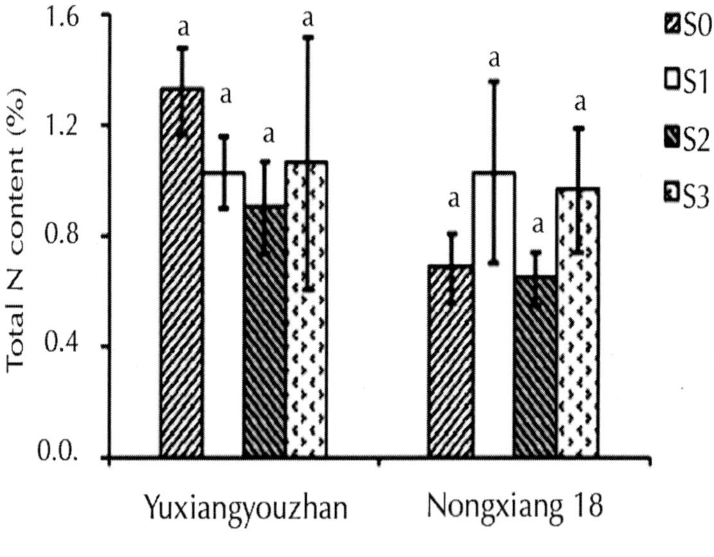

Figure 4: Effect of shading treatment on total nitrogen content in grains. Vertical bars with different lower case letters above are significantly different at $P=0.05$ by LSD tests. Capped bars represent SD.

Relationships Among 2-Acetyl-1-Pyrroline, GABA, Proline, and Total Nitrogen

Significant correlations were identified between 2-AP and GABA in response to shading for Yuxiangyouzhan but none for Nongxiang 18. No correlation was found between total nitrogen and GABA, proline or 2-AP and between proline and total nitrogen, GABA or 2-AP in either variety (Table 3).

Table 3: Correlation coefficients among 2-AP, GABA, total nitrogen, and proline content in grains

Index	2-AP content	Grain GABA content	Proline content	Total nitrogen content
Yuxiangyouzhan				
2-AP content	1	0.8326**	0.5460	−0.5557
GABA content	0.8326**	1	0.6926	−0.3402
Proline content	0.5460	0.6926	1	−0.2272
Total nitrogen content	−0.5557	−0.3402	−0.2272	1
Nongxiang 18				
2-AP content	1	0.5442	0.5356	−0.1758
GABA content	0.5442	1	0.3565	0.3836
Proline content	0.5356	0.3565	1	0.3267
Total nitrogen content	−0.1758	0.3836	0.3267	1

Significant correlations at $^{**}p<0.01$.

Mo *et al.*

Mo *et al.* *Rice* 2015 8:9, doi: 10.1186/s12284-015-0040-y

Table 4: Effect of shading treatment on the relative content (%) volatile compounds in grains

Compounds	Yuxiangyouzhan					Nongxiang 18				
	S0	**S1**	**S2**	**S3**	**Mean**	**S0**	**S1**	**S2**	**S3**	**Mean**
(E)-2-Hexenal	0.58c	1.76a	1.67a	1.17b	1.29	1.15b	2.16a	1.86a	1.55ab	1.68*
1-Hexanol	1.45a	1.12a	1.65a	1.69a	1.48	2.10ab	2.55a	2.21ab	1.78b	2.16
Heptanal	15.73a	19.84a	17.97a	16.28a	17.45	25.11ab	22.99b	24.14ab	25.91a	24.51*
2-acetyl-1-pyrroline	18.06b	18.07b	18.08b	22.93a	19.29*	10.84b	14.22a	14.17a	13.11a	13.08
Octane	7.26a	7.22a	5.91b	5.83b	6.55	7.00ab	7.47a	6.25bc	6.01c	6.68*
1-Heptanol	1.29a	1.07a	0.85a	0.86a	1.02	1.06a	0.98a	1.29a	1.02a	1.09
1-Octen-3-ol	1.96a	2.16a	3.43a	1.80a	2.34	2.12a	1.94a	1.87a	2.08a	2.00
Octanal	11.92b	14.81a	13.56ab	11.60b	12.97	13.36b	13.62b	13.26b	15.24a	13.87*
Benzyl alcohol	0.24b	0.74a	0.38ab	0.61ab	0.52	0.20b	0.78a	0.27b	0.82a	0.49
Benzeneacetaldehyde	5.31a	3.95a	4.00a	5.53a	4.70	4.16ab	4.90a	4.16ab	3.73b	4.23
3,8-Dimethylundecane	12.08a	9.62a	9.81a	13.36a	11.22	9.46b	12.16a	12.22a	8.67b	10.63

Means in the same row followed by different lower case letters for the same variety differ significantly at $P=0.05$ by LSD tests. Means of the two varieties followed by asterisk for the same compound difer significant at $P=0.05$ by LSD tests.

Mo *et al.*

Mo *et al.* *Rice* 2015 8:9, doi: 10.1186/s12284-015-0040-y

Effect of Shading Treatment on the Relative Content (%) Volatile Compounds

Eleven volatile compounds were measured: (E)-2-Hexenal, 1-Hexanol, Heptanal, 2-acetyl-1-pyrroline, Octane, 1-Heptanol, 1-Octen-3-ol, Octanal, Benzyl alcohol, Benzeneacetaldehyde, and 3, 8-Dimethylundecane. For Yuxiangyouzhan, shading treatments significantly increased relative content in (E)-2-Hexenal. S3 significantly increased relative content in 2-AP. shading treatments (S2 and S3) significantly decreased relative content in Octane. S1 treatment significantly increased the relative content of Octanal and Benzyl alcohol. For Nongxiang 18, shading treatments (S2 and S3) significantly increased the relative content of (E)-2-Hexenal. Shading treatments significantly increased the relative content of 2-AP. S3 significantly decreased the relative content of Octane, while significantly increased the relative content of Octanal. Shading treatment (S1 and S3) significantly increased the relative content of Benzyl alcohol. Shading treatments (S1and S2) significantly increased the relative content of 3,8-Dimethylundecane. As for the comparision between the two varieties, Yuxiangyouzhan had a significant higher relative content of 2-AP than Nongxiang 18. The relative content of (E)-2-Hexenal, Heptanal, Octane,and Octanal in Nongxiang 18 was significantly higher than that of Yuxiangyouzhan (Table 4).

DISCUSSION

Shading limits photosynthetically active radiation (PAR) and can be a source of plant stress. Several studies have assessed the effect of shading on growth and yield in cereal crops including rice; shading during flowering and grain development has been consistently found to decrease yield (e.g. Beed et al., [2007]; Zhang et al., [2006]; Liu and Tollenaar, [2009]; Deng et al., [2009]). Zhang et al. ([2007]) found that shading affects quality traits in rice, and generally, shading resulted in increased grain protein content and decreased amylose content, as well as an increased rate of chalky rice grains and a greater average degree of chalkinesss. In this study, during the whole (S1), early (S2), or late (S3) grain filling period, shading treatment resulted in significantly decreased yield in two elite Chinese fragrant rice varieties, with S1 treatment resulting in the greatest yield losses. Yield losses were more substantial in Yuxiangyouzhan than Nongxiang 18 under all shading treatments; this is consistent with previous studies in which genotype has been found to influence yield loss due to shading in rice (Zhu et al., [2008]; Deng et al., [2009]). Shading treatment also significantly affected quality traits. An increase in grain protein content under all treatments and a consistent effect of treatment on grain chalkiness were observed for both varieties. Significant shading effects were also identified for most other quality traits assessed, however these varied with timing, duration of treatment, and cultivar. Together, these results are largely consistent with previous literature and confirm that shading during grain development can have a substantial effect on grain yield and quality in fragrant rice.

Here we have found that all shading treatments significantly increased the concentration of both 2-AP and GABA in the grain of Yuxiangyouzhan than Nongxiang 18; furthermore, positive correlation between the accumulations of these compounds was observed in both varieties. Although GABA has been shown to accumulate in response to a range of stresses, direct assessment of the impact of shade on GABA accumulation appears limited; our results demonstrate that the conserved response of GABA to plant stress extends to shading in rice. Furthermore, our results indicate that the association between plant stress and accumulation of 2-AP exists for shading in addition to drought and salt. We also assessed accumulation of proline in reponse to shading treatment. This was significantly increased in Yuxiangyouzhan by S1 and S2 treatment; in contrast, no significant effect was detected for any shading treatment in Nongxiang 18. This suggests that there may be substantial variation of proline accumulation in response to shading in fragrant rice varieties.

Our observations of the accumulation of GABA and 2-AP in response to shading draw attention to the hypothesized relationship of the biochemistry of these compounds in fragrant rice. As outlined above,*in vitro* rice BADH2 catalyses the conversion of GABald to GABA, a molecule that has been widely implicated in plant stress responses (Bouche and Fromm, [2004]; Kinnersley and Turano, [2000]) and it has been hypothesized that in fragrant rice non-functional *BADH2* leads to the accumulation of GABald, which cyclises and reacts with acetyl group forming 2-AP (Bradbury et al., [2008]). However, although GABA can be synthesized in plants via GABald, the primary pathway for GABA production is thought to be the 'GABA shunt', in which glutamate is converted to GABA via glutamate decarboxylase (GAD) (Fait et al., [2008]).

If 2-AP indeed accumulates in fragrant rice via a pathway that produces GABA from GABald in the presence of functional *BADH2* (Bradbury et al., [2008]), it is intriguing that the accumulation of both GABA and 2-AP increases in response to shade. This suggests that there may be parallel activation of pathways leading to GABA in response to shading in fragrant rice, with 2-AP accumulating from GABald rather than GABA due to non-functional *BADH2*, but GABA also accumulating via the GABA shunt. However, it is possible that the accumulation of both GABA and 2-AP occurs via GABald. If another enzyme exists in rice that catalyses the conversion of GABald to GABA, non-functional BADH2 may result in a loss of efficiency rather than elimination of GABA accumulation via GABald, with excess GABald leading to accumulation of 2-AP. Bradbury et al., ([2008]) demonstrated that BADH1, an enzyme with ~ 76% identity to BADH2, catalyses the formation of GABA from GABald *in vitro*, albeit at lower efficiency than BADH2. Furthermore, Singh et al., ([2010]) identified an association of SNPs predicted to decrease activity of BADH1 on GABald with 2-AP-based aroma strength in rice. Thus, this latter possibility seems plausible. In either case, modulation of GABA accumulation in response to stress seems a credible explanation for the decrease in stress tolerance reported to be associated with fragrance in rice (Fitzgerald et al., [2010]; Wijerathna et al., [2014]).

For the other volatile compounds in grains, Buttery et al. ([1988]) demonstrated ten probable key contributors to cooked rice aroma among the detected compounds. Two of the ten volatile compounds, 2-acetyl-l-pyrroline and Octanal have been studied extensively (e.g. Mahatheeranont et al., [2001]; Jezussek et al., [2002]; Wongpornchai et al., [2003]; Laohakunjit and Noomhorm,[2004]; Yang et al., [2008]; Maraval et al., [2008]; Tananuwong and Lertsiri, [2010]; Goufo et al.,[2011]; Goufo et al., [2010]; Bryant and McClung, [2010]; Mathure et al., [2014]; Mahattanatawee and Rouseff, [2014]; Liyanaarachchi et al., [2014]). In our study, 2-AP had the highest average relative content among the detected compounds in Yuxiangyouzhan, while was the the third most abundant volatile detected in Nongxiang 18. Octanal had the second higher average relative content among the detected compounds in both varieties. There was a significant difference in the average average relative content of the two compounds between the two varieties. All shading treatment significantly increased the relative content in 2-AP in Nongxiang 18, while S3 had a significant impact on relative content in 2-AP in Yuxiangyouzhan. S1 significantly increased the relative content of octanal in both varieties. Of the compounds that were posited to contribute to fragrance in rice by Yang et al. ([2008]), in addition to 2-AP and octanal, heptanal and 1-octen-3-ol were also detected in our studies; and these two compounds have also been reported in other studies (Yang et al., [2008]; Tananuwong and Lertsiri, [2010]; Goufo et al., [2010]; Bryant and McClung, [2010]; Mathure et al.,[2014]). We found that, heptanal had the highest average relative content in Nongxiang 18, and had the second high average relative content in Yuxiangyouzhan. Nongxiang 18 had a significant higher average relative content of heptanal than Yuxiangyouzhan. For other volatile compounds, (E)-2-hexenal (Buttery et al.,[1988]; Yang et al., [2008]; Goufo et al., [2010] , [2011]), 1-hexanol (Mahatheeranont et al., [2001]; Goufo et al., [2010], [2011]; Mathure et al., [2014]), octane (Laohakunjit and Noomhorm, [2004]), 1-heptanol (Mahatheeranont et al., [2001]; Yang et al.,[2008]; Goufo et al., [2010], [2011]; Liyanaarachchi et al., [2014]), benzyl alcohol (Wongpornchai et al., [2003]; Mathure et al., [2014]; Liyanaarachchi et al., [2014]), benzeneacetaldehyde (Laohakunjit and Noomhorm, [2004]), and 3,8-dimethylundecane (Mahatheeranont et al., [2001]) were also detected in our study. Shading increased the relative content of (E)-2-hexenal in both varieties. There was no significant difference between shading and non-shading treatments in 1-hexanol, 1-heptanol, and benzeneacetaldehyde. Shading (S2 and S3 for Yuxiangyouzhan, S3 for Nongxiang 18) decreased the relative content of octane, and (S1 for Yuxiangyouzhan, S1 and S3 for Nongxiang 18) significantly increased the relative content of benzyl alcohol. The significant difference in 3,8-dimethylundecane was only found for Nongxiang 18 in S1 and S2 treatments. The results suggest that shading treatments may have a selective effect on the metabolism of the volatile compounds in fragrant rice.

Rice yield, grain quality and 2-AP content are key qualities for aromatic rice. In this study, we have assessed the impact of shading impact on aromatic rice yield, grain quality and 2-AP content, and some related traits. In addition to providing further insight into shading effects on crop yield and quality, and the interaction between 2-AP accumulation and stress in fragrant rice, the results of this study may have implications for rice cultivation in an environment being dramatically influenced by anthropogenic factors. Air pollution can have a shading effect, causing substantial decreases in surface level solar intensity (Jáuregui and Luyando, [1999]; Qian et al., [2006]), and China is well-known to experience extreme levels of air pollution (Chan and Yao, [2008]) including in major agricultural regions such as the Yangtze River delta (Shao et al., [2006]). There is evidence that the impact of air pollution has decreased annual solar radiation levels in Guangzhou and even the whole South China region (Liu et al., [2014]). Our results suggest that high pollution levels, particularly during grain filling, could influence yield, quality and aroma in elite Chinese fragrant rice varieties. While an increased content of 2-AP as a result of the shading effect of air pollution may be desirable, yield loss and loss of grain quality would clearly be detrimental.

CONCLUSIONS

Here, we have demonstrated that shading during grain filling has significant effects on yield and quality traits in rice, and leads to the accumulation of GABA and 2-AP. These results further emphasize the important relationship between stress and 2-AP production in fragrant rice. Continuing advances in the understanding of this relationship have great potential for the optimization of both aroma production and stress resistance in elite rice varieties.

MATERIALS AND METHODS

Plant Materials and Treatment Conditions

Two aromatic rice cultivars, Yuxiangyouzhan and Nongxiang 18 were used in this study. These are the main commercial fragrant rice cultivars in South China. Field experiment was carried out from July to November, 2013 at South China Agricultural University's Experimental Farm. This region has a humid subtropical monsoonal climate. The field consists of sandy, loamy soil and has been under paddy cultivation for many years. The main soil properties of the experimental site were as follows:

pH, 4.88; organic matter content, 25.65%; total N, 1.362%; total P, 0.958%; total K, 17.520%.

The experiment was arranged in split-plot design with varieties as the main plots and shading treatment as the subplots with three duplications. After years of planting, it is well-known that the duration of grain filling in the two varieties employed for this study is approximaltey 30 days. Based on previous studies, we found that a very strong shading level, corresponding to a 90% reduction in full natural light (as measured by a Luxmeter, model ZDS-10, China) resulted in a substantial increase in pest and diseases damage. Therefore, we employed a shading level equivalent to a 67% reduction of full natural light (as measured by a, Luxmeter, model ZDS-10, China) created by one layer of black netting. Four shading treatments were studied: (i) S0: the whole phase during grain filling without shading, taken as the control; (ii) S1: shading during the whole phase of grain filling (30 days, from October 3rd to November 2nd); (iii) S2: shading during the early phase of grain filling (15 days, from October 3rd to October 18th); (iv) S3: shading during the later phase of the grain filling (15 days, from October 18th to November 2nd). The whole grain filling stage was from the end of pollination to grain dry-down that is from R5 to R9 in the report Counce et al. ([2000]), equivalent to about 30 days for the two experimental varieties. As outlined above, the whole grain filling stage was devided into two stages, one was the early phase of grain filling (15 days), the other was the later phase of grain filling (15 days).

The area for each subplot was 16 m^2. Seeds were sown on July 15th and transplanting was performed on August 7th at a density of 20 cm × 20 cm (about 400 hills per plot) with two seedlings per hill. Plants were harvested on November 5th. Fertilizer (1500 $kg{\cdot}ha^{-1}$; N, 12.5%; P_2O_5, 6.0%; K_2O, 10.0%; organic matter, 15.0%) was applied 60% at basal, 40% at tillering stage. Irrigation, pest and diseases management, and weed control were the same in all treatments following the guidelines recommended by the province.

Sampling and Measurement

Determination of Dry Weight

Plants from ten randomly selected hills were taken to the laboratory at maturity. Plants were separated into leaves, (stem + sheath) and panicles. Sample was then oven-dried at 80°C to constant weight for the determination of dry weight.

Determination of Total Nitrogen Content

Total nitrogen content in grain was determined as described by Lu ([1999]). Briefly, the dried samples of ground grains (about 0.3 g) were digested using the H_2SO_4-$HClO_4$ method. The digestion was then used to determine the total nitrogen content by the Kjeldahl method with a 2300 Kjeltec Analyzer Unit (Foss Tecator AB, Sweden).

Determination of Yield and Yield Related Traits

At maturity, grain yield was measured from one unit sampling area (1 m^2) within each plot, threshed manually, then sun dried (adjusted to moisture content of ~ 14%). Plants from ten randomly selected hills from each plot were sampled to investigate yield-related traits. Panicle number per m^2 was measured by counting the panicle numbers of each hill within the unit area (1 m^2) at three different locations in each plot and the mean value was taken as the final result. Panicles were threshed manually, total number of grains and number of filled grains were counted. Five samples of 1000 grains were taken randomly from filled seeds, weighted to record 1000-grain weight. Harvest index was calculated as the ratio between grain yield and dry weight at maturity, expressed as a percentage.

Determination of Grain Quality

About 1 kg rice grain from each treatment was obtained after storage at room temperature for three months. Brown rice rate was estimated using a rice huller (Jiangsu, China). Milled rice and head rice rates were estimated using a Jingmi testing rice grader (Zhejiang, China). Percentage of grain with chalkiness and chalkiness degree were estimated using an SDE-A light box (Guangzhou, China). Amylose content, protein content, and alkali of grains were determined using an Infratec 1241 grain analyzer (FOSS-TECATOR).

Determination of Proline Content, Gaba Content and 2-Ap Concentration

For determination of proline, GABA, and 2-AP, fresh sample of grains from 1 m^2 area of each plot were harvested and immediately stored at −20°C. Duplicate measurements were taken for each sample and the mean value of the three samples per treatment was taken as the final result.

The proline content was determined according to the method described by Bates et al. ([1973]). Briefly, grains (about 0.3 g) was homogenized in 5 mL of 3% sulfosalicylic acid, then cooled after heating at boiling water bath for 10 min. Samples filtered and two ml of the filtrate was mixed with 3 mL of ninhydrin reagent (1.25 g ninhydrin in 30 mL glacial acetic acid and 20 mL 6 M phosphoric acid) and 2 mL of glacial acetic acid. The reaction mixture was then heated at boiling water bath for 30 min and placed in an ice bath for 20 min before being extracted with 4 ml of toluene. The toluene extraction was then centrifuged at 4000 rpm for 5 min. The absorbance of the red chromophore in the toluene fraction was measured at 520 nm and the amount of proline was determined by comparison with a standard curve and expressed as $\mu g \cdot g^{-1}$.

GABA content in grain was determined as described by Zhao et al. ([2009]) and Yao et al. ([2008]). Briefly, grain (about 0.5 g) was homogenized in 5 mL of 60% ethanol, treated for 4 hours in a oscillations instrument (HZS-H, China) using a frequency of 200 oscillations per minute. The supernatant was then transferred to a 5 ml centrifuge tube and centrifuged at 8000 rpm for 3 min. 1 mL of supernatant was added to a 10 mL tube, mixed with 0.6 mL 0.2 $mol \cdot L^{-1}$ (pH 9.0) sodium tetraborate, two mL 5% toluene, and 1 mL 7% sodium hypochlorite, then cooled after heating at 100°C in a water bath for 5 minutes. The absorbance of the reaction solution was measured at 645 nm and the amount of GABA was determined by comparison with a standard curve and expressed as μg·g^{-1}.

Grain samples were evaluated for 2-AP concentration by synchronization distillation and extraction method (SDE) combined with GCMS-QP 2010 Plus (Shimadzu Corporation, Japan) as described by Huang et al. ([2012]). Briefly, Collidine (2, 4, 6-trimethylpyridine) (Sigma, Switzerland), was used as an internal standard. 10 g of finely ground grain was transferred into a 500 mL round-bottom flask containing 145 mL purified water; 5 mL of 0.914 μg·mL^{-1} internal standard was then added. A steam distillation continuous extraction head was attached to the flask and the flask was heated at 150°C by an oil bath pot, ZKYY (Guangzhou, China). Attached to the other head of the steam distillation continuous extraction instrument, diethyl ether (35 mL) was used as the solvent in a 500 mL round-bottom flask, and the flask was heated at 42°C by a water bath pot, HH-2 (Jiangsu, China). During isolation, the steam distillation continuous extraction was maintained at 10°C by a cold water circulation machine, YKKY-LX-300 (Beijing, China). The isolation was performed for 35 min. The ether extract was then dried over sodium sulfate, filtered (0.22 μm filter paper, Shimadzu, Japan), and then directly used to measure 2-AP concentrations with the GCMS-QP 2010 Plus method as described in Huang et al., ([2012]). The GCMS-QP 2010 Plus working conditions were as followings: gas chromatograph equipped with a Restek Rxi-5 ms (Shimadzu, Japan) silica capillary column (30 m×0. 32 mm×0. 25 μm). The auto injector was AOC-20i, SPL1. High purity helium gas (99.999%, Guangzhou Gases Co., LTD, China) was the carrier gas at the flow rate of 2.0 mL·min^{-1}. The temperature of the GC oven was 40°C (1 min), increased at 2°C·min^{-1} to 65°C and held at 65°C for 1 min, and then increased to 220°C at 10°C·min^{-1}, and held at 220°C for 10 min. The ion source temperature was 200°C. Under these conditions, the retention time of 2-AP was 7.5 min. 2-AP content was expressed as μg·kg^{-1}. The relative content of the aroma compounds were identified on the basis of their mass spectra by comparing the spectra with the records of the NIST library.

Statistical Analysis

Analysis of variance and correlation coefficients were performed using Statistix 8 (Analytical, Tallahassee, Florida, USA). The data were analyzed by one-way analysis of variance to assess differences in yield, yield related traits, total dry weight, harvest index, grain quality, 2-acetyl-1-pyrroline, GABA, proline and total nitrogen between treatments.

AUTHORS' CONTRIBUTIONS

ZM, XT, WL and SP designed the experiments, ZM,WL and TLF analyzed the data, and wrote the article, FX,YT, YW and HT performed the traits investigation, ZM and MD conducted the GC-MS analysis. All authors read and approved the final manuscript.

ACKNOWLEDGEMENTS

This study was supported by National Natural Science Foundation of China (31271646), Natural Science Foundation of Guangdong Province (8151064201000017), Agricultural Research Projects of Guangdong Province (2011AO20202001), and Agricultural Standardization Project of Guangdong Province (4100 F10003). The authors would like to thank Dr. Aryadeep Roychoudhury and Altafhusain B Nadaf for their contributions to the manuscript.

REFERENCES

1. Bates LS, Waldron RP, Teare ID: Rapid determination of free proline for water stress studies. *Plant and Soil* 1973, 39:205-208.
2. Beed FD, Paveley ND, Sylvester-Bradley R: Predictability of wheat growth and yield in light-limited conditions. *J Agric Sci* 2007, 145(01):63-79.
3. Bouche N, Fromm H: GABA in plants: just a metabolite? *Trends Plant Sci* 2004, 9(3):110-115.

4. Bradbury LMT, Fitzgerald TL, Henry RJ, Jin Q, Waters DL: The gene for fragrance in rice. *Plant Biotechnol J* 2005, 3(3):363-370.
5. Bradbury LM, Gillies SA, Brushett DJ, Waters DL, Henry RJ: Inactivation of an aminoaldehyde dehydrogenase is responsible for fragrance in rice. *Plant Mol Biol* 2008, 68(4–5):439-449.
6. Bryant RJ, McClung AM: Volatile profiles of aromatic and non-aromatic rice cultivars using SPME/GC–MS. *Food Chem* 2010, 124(2):501-513.
7. Buttery RG, Ling LC, Mon TR: Quantitative analysis of 2-acetyl-1-pyrroline in rice. *J Agric Food Chem* 1986, 34(1):112-114.
8. Buttery RG, Turnbaugh JG, Ling LC: Contribution of volatiles to rice aroma. *J Agric Food Chem* 1988, 36(5):1006-1009.
9. Buttery RG, Ling LC, Juliano BO, Turnbaugh JG: Cooked rice aroma and 2-acetyl-1-pyrroline. *J Agric Food Chem* 1983, 31(4):823-826.
10. Cai KZ, Luo SM: Effect of shading on growth, development and yield formation of rice. *Chinese Journal of Applied Ecology* 1999, 10(2):193-196. (in Chinese)
11. Champagne ET: Rice aroma and flavor: a literature review. *Cereal Chemistry* 2008, 85(4):445-454.
12. Champagne ET, Bett-Garber KL, Thompson J, Mutters R, Grimm CC, McClung AM: Effects of drain and harvest dates on rice sensory and physicochemical properties. *Cereal Chemistry* 2005, 82(4):369-374.
13. Chan CK, Yao X: Air pollution in mega cities in China. *Atmos Environ* 2008, 42(1):1-42.
14. Counce PA, Keisling TC, Mitchell AJ: A uniform, objective, and adaptive system for expressing rice development. *Crop Sci* 2000, 40(2):436-443.
15. Deng F, Wang L, Yao X, Wang JJ, Ren WJ, Yang WY: Effects of different-growing-stage shading on rice grain-filling and yield. *Journal of Sichuan Agricultural University* 2009, 27(3):265-269. (in Chinese)
16. Ding SB, Zhu BY, Wu DY, Zhang L: Effect of temperature and light on senescence of flag leaf and grain-filling after rice heading.

Journal of South China Normal University (Natural Science Edition) 2004, 1:117-121,128. (in Chinese)

17. Fait A, Fromm H, Walter D, Galili G, Fernie AR: Highway or byway: the metabolic role of the GABA shunt in plants. *Trends Plant Sci* 2008, 13(1):14-19.
18. Fitzgerald TL, Waters DLE, Brooks LO, Henry RJ: Fragrance in rice (*Oryza sativa*) is associated with reduced yield under salt treatment. *Environmental and Experimental Botany* 2010, 68(3):292-300.
19. Gay F, Maraval I, Roques S, Gunata Z, Boulanger R, Audebert A, Mestres C: Effect of salinity on yield and 2-acetyl-1-pyrroline content in the grains of three fragrant rice cultivars (*Oryza sativa* L.) in Camargue (France). *Field Crop Res* 2010, 117(1):154-160.
20. Gommers CM, Visser EJ, Onge KRS, Voesenek LA, Pierik R: Shade tolerance: when growing tall is not an option. *Trends Plant Sci* 2013, 18(2):65-71.
21. Goufo P, Duan M, Wongpornchai S, Tang X: Some factors affecting the concentration of the aroma compound 2-acetyl-1-pyrroline in two fragrant rice cultivars grown in South China. *Frontiers of Agriculture in China* 2010, 4(1):1-9.
22. Goufo P, Wongpornchai S, Tang X: Decrease in rice aroma after application of growth regulators. *Agronomy for Sustainable Development* 2011, 31:349-359.
23. Hashemi FSG, Rafii MY, Ismail MR, Mahmud TMM, Rahim HA, Asfaliza R, Malek MA, Latif MA:Biochemical, genetic and molecular advances of fragrance characteristics in rice. *Crit Rev Plant Sci* 2013, 32(6):445-457.
24. Haywood J M, Bellouin N, Jones A, Boucher O, Wild M, Shine K P (2011) The roles of aerosol, water vapor and cloud in future global dimming/brightening. Journal of Geophysical Research: Atmospheres (1984–2012), 116(D20203), doi:10.1029/2011JD016000
25. Huang ZL, Tang XR, Wang YL, Chen MJ, Zhao ZK, Duan MY, Pan SG: Effects of increasing aroma cultivation on aroma and grain yield of aromatic rice and their mechanism. *Scientia Agricultura Sinica* 2012, 45(6):1054-1065. (in Chinese)

26. Jáuregui E, Luyando E: Global radiation attenuation by air pollution and its effects on the thermal climate in Mexico City. *Int J Climatol* 1999, 19(6):683-694.
27. Jezussek M, Juliano BO, Schieberle P: Comparison of key aroma compounds in cooked brown rice varieties based on aroma extract dilution analyses. *J Agric Food Chem* 2002, 50(5):1101-1105.
28. Kinnersley AM, Turano FJ: Gamma aminobutyric acid (GABA) and plant responses to stress. *Crit Rev Plant Sci* 2000, 19(6):479-509.
29. Laohakunjit N, Noomhorm A: Supercritical carbon dioxide extraction of 2-acetyl-1-pyrroline and volatile components from pandan leaves. *Flavour and Fragrance Journal* 2004, 19(3):251-259.
30. Liu AJ, Chen WC, Qin Q: An Analysis and estimation on Solar energy resources in Guangzhou city.Guangdong. *Meteorology* 2014, 36(3):59-61. (In Chinese)
31. Liu B, Han Y, Xie WX, Li JG, Liu J, Gao Q: The impact of weak light during grain filling on rice yield, physiology and quality. *Chinese Rice* 2008, 5:36-40. (in Chinese)
32. Liu L, Wang L, Deng F, Huang Y, Liu DY, Ren WJ, Yang WY: Osmotic regulation substance content and activities of protective enzymes in leaves of diffenent hybrid rice combinations as affected by shading. *Chinese Journal of Rice Science* 2012, 26(5):569-575. (in Chinese)
33. Liu W, Tollenaar M: Response of yield heterosis to increasing plant density in maize. *Crop Sci* 2009, 49(5):1807-1816.
34. Liyanaarachchi GD, Kottearachchi NS, Samarasekera R: Volatile profiles of traditional aromatic rice varieties in Sri Lanka. *Journal of the National Science Foundation of Sri Lanka* 2014, 42(1):87-93.
35. Lu RK: *Analys is measures for agro-chemical on soil.* China Agricultural Science and Technology Press, Beijing; 1999.
36. Mahatheeranont S, Keawsa-ard S, Dumri K: Quantification of the rice aroma compound, 2-acetyl-1-pyrroline, in uncooked Khao Dawk Mali 105 brown rice. *J Agric Food Chem* 2001, 49(2):773-779.

37. Mahattanatawee K, Rouseff RL: Comparison of aroma active and sulfur volatiles in three fragrant rice cultivars using GC–Olfactometry and GC–PFPD. *Food Chem* 2014, 154:1-6.
38. Maraval I, Mestres C, Pernin K, Ribeyre F, Boulanger R, Guichard E, Gunata Z: Odor-active compounds in cooked rice cultivars from Camargue (France) analyzed by GC– O and GC– MS. *J Agric Food Chem* 2008, 56(13):5291-5298.
39. Maraval I, Sen K, Agrebi A, Menut C, Morere A, Boulanger R, Gay F, Mestresa C, Gunata Z: Quantification of 2-acetyl-1-pyrroline in rice by stable isotope dilution assay through headspace solid-phase microextraction coupled to gas chromatography–tandem mass spectrometry. *Anal Chim Acta* 2010, 675(2):148-155.
40. Mathure SV, Jawali N, Thengane RJ, Nadaf AB: Comparative quantitative analysis of headspace volatiles and their association with *BADH2* marker in non-basmati scented, basmati and non-scented rice (*Oryza sativa* L.) cultivars of India. *Food Chem* 2014, 142:383-391.
41. McCree KJ (1981) Photosynthetically active radiation. Physiological Plant Ecology I, Springer Berlin Heidelberg 41–55
42. Niu XL, Tang W, Huang WZ, Ren GJ, Wang QL, Luo D, Xiao YY, Yang SM, Wang F, Lu BR, Gao FY, Lu TG, Liu YS: RNAi-directed downregulation of *OsBADH2* results in aroma (2-acetyl-1-pyrroline) production in rice (*Oryza sativa* L.). *BMC Plant Biol* 2008, 8(1):100
43. Poonlaphdecha J, Maraval I, Roques S, Audebert A, Boulanger R, Bry X, Gunata Z: Effect of timing and duration of salt treatment during growth of a fragrant rice variety on yield and 2-Acetyl-1-pyrroline, proline, and GABA levels. *J Agric Food Chem* 2012, 60(15):3824-3830.
44. Qian Y, Kaiser DP, Leung LR, Xu M: More frequent cloud-free sky and less surface solar radiation in China from 1955 to 2000. *Geophys Res Lett* 2006., 33(1)
45. Ren WJ, Yang WY, Fan GQ, Zhu XM, Ma ZH, Xu JW: Effect of low light on dry matter accumulation and yield of rice. *Journal of Sichuan Agricultural University* 2003, 21(4):292-296. (in Chinese)
46. Sakthivel K, Sundaram RM, Shobha Rani N, Balachandran SM, Neeraja CN: Genetic and molecular basis of fragrance in rice. *Biotechnol Adv* 2009, 27(4):468-473.

47. Shao M, Tang XY, Zhang YH, Li WJ: City clusters in China: air and surface water pollution. *Front Ecol Environ* 2006, 4(7):353-361.

48. Singh A, Singh PK, Singh R, Pandit A, Mahato AK, Gupta DK, Tyagi K, Singh AK, Sharma TR:SNP haplotypes of the *BADH1* gene and their association with aroma in rice (*Oryza sativa* L.). *Molecular breeding* 2010, 26(2):325-338.

49. Szabados L, Savouré A: Proline: a multifunctional amino acid. *Trends Plant Sci* 2010, 15(2):89-97.

50. Tanaka A, Kawano K: Effect of mutual shading on dry-matter production in the tropical rice plant. *Plant and Soil* 1966, 24(1):128-144.

51. Tananuwong K, Lertsiri S: Changes in volatile aroma compounds of organic fragrant rice during storage under different conditions. *J Sci Food Agric* 2010, 90(10):1590-1596.

52. Tang YY: Effects of shading on growth, development and yield-constituting factors of rice. *Plant Physiology Communications* 1988, 2:50-53. (in Chinese)

53. Tsugita T: Aroma of cooked rice. *Food Reviews International* 1985, 1(3):497-520.

54. Wang L, Deng F, Ren WJ, Yang WY: Effects of Shading on starch pasting characteristics of Indica hybrid rice (*Oryza sativa* L.). *PLoS One* 2013, 8(7):e68220.

55. Widjaja R, Craske JD, Wootton M: Comparative studies on volatile components of non-fragrant and fragrant rices. *J Sci Food Agric* 1996, 70(2):151-161.

56. Wijerathna YMAM, Kottearachchi NS, Gimhani DR, Sirisena DN: Exploration of relationship between fragrant gene and growth performances of fragrant rice (*Oryza sativa* L.) seedlings under salinity stress. *Journal of Experimental Biology and Agricultural Sciences* 2014, 2(1):7-12.

57. Wongpornchai S, Sriseadka T, Choonvisase S: Identification and quantitation of the rice aroma compound, 2-acetyl-1-pyrroline, in bread flowers (Vallaris glabra Ktze). *J Agric Food Chem* 2003, 51(2):457-462.

58. Yang DS, Shewfelt RL, Lee KS, Kays SJ: Comparison of odor-active compounds from six distinctly different rice flavor types. *J Agric Food Chem* 2008, 56(8):2780-2787.

59. Yao S, Yang TW, Zhao LJ, Xiong SB: The variation of γ-aminobutyric acid content in germinated brown rice among different cultivars. *Scientia Agricultura Sinica* 2008, 41(12):3974-3982. (in Chinese)

60. Yoshihashi T, Kabaki N, Nguyen TTH, Inatomi H: Formation of flavor compound in aromatic rice and its fluctuations with drought stress. *Research Highlights JIRCAS* 2002, 2003(2004):32-33.

61. Zhang JW, Dong ST, Wang KJ, Hu CH, Liu P: Effects of shading on the growth, development and grain yield of summer maize. *Chinese Journal of Applied Ecology* 2006, 17(4):657-662. (in Chinese)

62. Zhang LL, ZhangWZ, Han YD, Wang ZH, Yan LX, Gao SK (2007) Effect of shding on rice quality of diffenent panicle types.Liaoning Agricultural Science (2): 18–21 (in Chinese)

63. Zhao DW, Pu XY, Zeng YW, Li BX, Du J, Yang SM: Determination of the γ-aminobutyric acid in barley. *Journal of Triticeae Crops* 2009, 29(1):69-72. (in Chinese)

64. Zhu P, Yang SM, Ma J, Li SX, Chen Y: Effect of shading on the photosynthetic characteristics and yield at later growth stage of hybrid rice combination. *Acta Agromomica Sinica* 2008, 34(11):2003-2009. (in Chinese)

Chapter 7

Pseudo-Backcrossing Design for Rapidly Pyramiding Multiple Traits Into a Preferential Rice Variety

Siriphat Ruengphayak[1,2], Ekawat Chaichumpoo[1], Supaporn Phromphan[1], Wintai Kamolsukyunyong[3], Wissarut Sukhaket[1], Ekapol Phuvanartnarubal[1], Siripar Korinsak[3],Siriporn Korinsak[3], and Apichart Vanavichit [1,3,4]

[1]Rice Science Center, Kasetsart University, Kamphaeng Saen 73140, Nakhon Pathom, Thailand

[2]Interdisciplinary Graduate Program in Genetic Engineering, Kasetsart University, Chatuchak 10900, Bangkok, Thailand

[3]Rice Gene Discovery, National Center for Genetic Engineering and Biotechnology (BIOTEC) National Science and Technology Development Agency (NSTDA), Kasetsart University, Kamphaengsaen 73140, Nakhon Pathom, Thailand

[4]Agronomy Department, Faculty of Agriculture at Kamphaeng Saen, Kasetsart University, Kamphaeng Saen 73140, Nakhon Pathom, Thailand

ABSTRACT

Background

Pyramiding multiple genes into a desirable genetic background can take years to accomplish. In this paper, a pseudo-backcrossing scheme was designed to shorten the backcrossing cycle needed. PinK3, an aromatic and potentially high-yielding rice variety—although one that is intolerant to flash flooding (Sub) and susceptible to bacterial leaf blight (BB), leaf-neck blast (BL) and the brown planthopper (BPH)—was used as a genetic basis for significant improvements through gene pyramiding.

Results

Four resistance donors with five target genes (*Sub1A-C, xa5, Xa21, TPS* and *SSIIa*) and three QTLs (qBph3, qBL1 and qBL11) were backcrossed individually using markers into the pseudo-recurrent parent 'PinK3' via one cycle of backcrossing followed by two cycles of pseudo-backcrossing and three selfings with rigorous foreground marker-assisted selection. In total, 29 pseudo-backcross inbred lines (BILs) were developed. Genome composition was surveyed using 61 simple sequence repeats (SSRs), 35 of which were located on six carrier chromosomes, with the remainder located on six non-carrier chromosomes. The recurrent genome content (%RGC) and donor genome content (%DGC), which were based on the physical positions of BC_1F_2, ranged from 69.99 to 88.98% and 11.02 to 30.01%, respectively. For the pseudo-BC_3F_3BILs, the %RGC and %DGC ranged from 74.50 to 81.30% and 18.70 to 25.50%, respectively. These results indicated that without direct background selection, no further increases in %RGC were obtained during pseudo-backcrossing, whereas rigorous foreground marker-assisted selection tended to reduce linkage drag during pseudo-backcrossing. The evaluation of new traits in selected pseudo-BC_3F_3BILs indicated significant improvements in resistance to BB, BL, BPH and Sub compared with PinK3, as well as significant improvements in grain yield (21-68%) over the donors, although yield was 7-26% lower than in 'PinK3'. All pyramided lines were aromatic and exhibited improved starch profiles, rendering them suitable for industrial food applications.

Conclusions

Results show that our new pyramiding platform, which is based on marker-assisted pseudo-backcrossing, can fix five target genes and three QTLs into a high-yielding pseudo-recurrent background within seven breeding cycles in four years. This multiple pseudo-backcrossing platform decreases the time required to generate new rice varieties exhibiting complex, durable resistance to biotic and abiotic stresses in backgrounds with desirable qualities.

BACKGROUND

Rice (*Oryza sativa* L.) is an important staple food crop and a major part of the diet of more than half of the world's population. Approximately 90% of rice is grown in Asia. Worldwide, approximately 79 million ha of irrigated lowland rice provides 75% of the world's rice production (Maclean et al. [2002]; Bouman et al. [2007]). Therefore, irrigated rice remains the most important production system for maintaining food security, particularly in Asian countries. Rice production in irrigated areas of Thailand has been frequently and strongly affected by abiotic stresses resulting from unfavorable climatic changes, such as flooding and drought, as well as by biotic stresses caused by bacterial leaf blight (BB), leaf/neck blast (BL) and brown planthopper (BPH). Therefore, new successful breeding lines must possess multiple types of resistance to both biotic and abiotic stresses, as well as demonstrating specific grain qualities and high yield.

Three popular breeding methods used for gene pyramiding are pedigree, backcrossing and recurrent selection. In cross-pollinating crops, gene pyramiding is accomplished through recurrent selection. Successful quantitative trait locus (QTL) introgression depends on the optimized expression of newly introgressed QTLs in the recipient genome background, with the aim of maximizing productivity. A general framework for addressing these considerations through the pyramiding of multiple QTLs into a single favorable genetic background has been proposed, although such techniques are time-consuming (Servin et al. [2004]).

To pyramid several new QTLs, stepwise crossing schemes can be designed, although such schemes can require many generations of breeding. If the ultimate goal is to improve a specific desired variety, parallel backcrossing of single donors can most effectively be carried out using both foreground and background selection. Recently, successful gene/QTL pyramiding programs were reported in Thai Jasmine (Win et al. [2012]; Luo and Yin, [2013]), Basmati (Singh et al.[2012]; Singh et al. [2013]), Koshihikari (Ashikari and Matsuoka, [2006]; Tomita, [2009]), Zhenshan 97 (Wang et al. [2012]) and 93–11 (Zong et al. [2012]) rice varieties. However, when the target traits are quantitatively controlled, combining several QTLs can take years to accomplish. For multiple QTL pyramiding, three phases are necessary: the creation of near-isogenic lines (NILs), genotype assembly, and the extraction of pure lines. Specific backcrossed recombinant inbred lines (RILs) carrying four main-effect QTLs and four epistatic-effect QTLs were pyramided into the elite cultivar Zhenshan97 (Wang et al. [2012]). To combine greater numbers of QTLs, marker-assisted phenotypic selection (MAS) has been developed, which is a novel approach for QTL pyramiding of up to 24 QTLs from a single crossing (Zong et al. [2012]). QTL pyramiding via NILs was successfully used to improve disease and lodging resistance, as well as to increase the harvest index (Luo and Yin, [2013]). However, all of these approaches require many years to complete.

To shorten the backcross breeding cycle, we propose a modified form of pseudo-backcrossing. The original design of pseudo-backcrossing originated from tree breeding methods, where F_1 plants resulting from a single cross are backcrossed to alternate recurrent parents to avoid inbreeding depression (Bouquet [1986]). Pseudo-backcrossing is commonly used in perennial plants, including grape (Molnár et al. [2007]), eucalyptus (Kullan et al. [2012]), poplar (Novaes et al. [2009]) and oil palm (Montoya et al. [2014]). For multiple gene/QTL pyramiding, several donors are used to create newly improved genotypes; however, the need to maintain the preferred genetic background is equally important. Therefore, the introduction of pseudo-backcrossing could benefit multiplex gene pyramiding.

Most rice varieties are not tolerant to flash flooding. Submergence tolerance is determined by the major Sub1QTL on chromosome 9 with relatively high heritability. The new Thai jasmine rice variety 'Khao Dawk Mali 105 (KDML105)' was developed for submergence (Sub) tolerance using marker-assisted backcross breeding (MAB) and has been released to farmers (Siangliw et al.[2003]). Growing rice plants also suffer from epidemic diseases such as rice blast (BL) and bacterial leaf blight (BB). Rice BL, which is caused by the fungal pathogen *Pyricularia oryzae*(teleomorph: *Magnaporthe oryzae*), is a major rice disease in irrigated rice-growing areas worldwide (Ou [1985]). QTLs for broad-spectrum resistance to rice BL have been reported on chromosomes 1 and 11 from JHN and on chromosomes 2 and 12 from IR64 (Sirithunya et al.[2002]; Noenplab et al. [2006]; Sreewongchai et al. [2010]). Major effective QTLs for Thai blast isolates were located on chromosomes 1 and 11, which are flanked by RM212 and RM319 on qBL1 and by RM224-RM144 on qBL11 (Noenplab et al. [2006]; Wongsaprom et al. [2010]). The presence of qBL1 and qBL11 have strong effects on blast resistance in Thailand. BB, which is caused by the bacterium *Xanthomonas oryzae* pv oryzae can be effectively controlled using resistant varieties. Several resistance genes, including *Xa4, xa5* and *Xa21,* are effective sources of resistance for marker-assisted gene pyramiding in rice (Korinsak et al. [2009b]; Suh et al. [2013]).

Among insect rice pests, the brown planthopper (BPH, *Nilaparvata lugens* Sta° l), is considered one of the most serious pests of irrigated rice. The damage caused by BPH feeding has a major effect on crop growth and yield (Watanabe and Kitagawa, [2000]; Yuan et al. [2005]). BPH not only feeds on the rice plant directly but also transmits viruses that cause severe diseases (Heinrichs [1979]). The use of BPH resistance genes has been recognized as the most economic, effective and environmentally friendly solution to this problem. The stability of BPH resistance in Rathu Heenati (RH), a traditional Sri Lankan rice cultivar containing qBph3, has made this strain one of the most popular hopper-resistance donors in the Mekong sub-region, where rice production is highly intensive. QTL mapping located qBph3 to the short arm of chromosome 6 between RM589 and RM588 based on KDML105 × Rathu Heenati (Jairin et al. [2009]) and PTB33 × RD6 (Jairin et al.[2007]) crossings. Other BPH resistance QTLs have also been reported, including Bph17 on chromosome 4 (Sun et al. [2005]), Bph4 on chromosome 6 (Kawaguchi et al. [2001]; Sun et al.[2006]) and Bph18 on chromosome 12 (Jena et al. [2006]). In addition to resistance QTLs, a putative sesquiterpene synthase (*TPS*) gene (Os04g27430) was identified using SFP mapping with isogenic lines derived from the backcross of RH and KDML105 (Kamolsukyunyong et al. [2013]). The *TPS* gene is induced after 5 days of BPH feeding. Functional markers were identified in exon 5 of the *TPS* gene that resulted in the deletion of seven amino acids in the susceptible rice line, as well as three additional SNPs associated with a transcriptional binding site, accounting for the differential response of *TPS* during the anti-feeding test (Kamolsukyunyong et al. [2013]).

Alkali disintegration has been used as a biomarker for gelatinization temperature (GT) in rice (Waters et al. [2006]; Kate-ngam et al. [2008]; Masouleh et al. [2012]). The alkali disintegration locus (*ALK*) was identified as the *starch synthase IIa* (*SSIIa*) gene on chromosome 6 that determines amylopectin structure (Bao et al. [2004]; Bao et al. [2006a]; Umemoto et al. [2002]; Umemoto et al. [2004]; Umemoto and Aoki, [2005]; Waters et al. [2006]; Lu et al. [2010]). Two functional SNPs were identified: GC/TT and G/A at positions 4329/4330 bp and 4198 bp, respectively (accession AY423717) (Bao et al. [2006b]; Umemoto et al. [2002]; Umemoto et al.[2004]; Umemoto and Aoki, [2005]; Waters et al. [2006]; Lu et al. [2010]). Successful marker assisted selection (MAS) programs utilizing the GT haplotypes for improving starch profiles have been reported (Tian et al. [2009]; Lu et al. [2010]).

'PinK3' is an aromatic, high-yielding, non-photoperiod-sensitive, high-amylose rice variety, but it is susceptible to BPH, BB, BL and Sub stresses. Here, we report the successful gene/QTL pyramiding of five functional genes (*xa5, Xa21, Sub1A-C, SSIIa, TPS*) and three QTLs (qBph3, qBL1, qBL11) into the 'PinK3' genome background using a multiplex pseudo-backcrossing approach based on MAS. The new, improved lines have a high-yield phenotype that confers submergence tolerance and resistance to BPH, BB and BL. This is the first report describing the application of pseudo-backcrossing to significantly shorten the time required for gene/QTL pyramiding in an annual crop (rice).

RESULTS

Streamline Gene Pyramiding in Rice

Four sets of donors (Table 1) containing a set of target genes/ QTLs—Cholsub (*Sub1A-C* and*SSIIa*), Xa497 (*xa5* and *Xa21*), Bph162 (qBph3 and *TPS*), and RBPiQ (qBL1 and qBL11)—were backcrossed in parallel once to 'PinK3' by targeting MAB to generate four sets of BC_1F_1 progeny. The resulting BC_1F_1 progeny were stepwise crossed to recombine four sets of the target genes/QTLs into a single set of pseudo-backcrossed progenies. By pair-wise crossing two pairs of BC_1F_1 lines, two sets of pseudo-backcrossed progeny—BC_2F_1 (*Sub1A-C, SSIIa, xa5* and *Xa21*) and BC_2F_1 (qBph3, *TPS*, qBL1 and qBL11)—were generated (Figure 1). In cycle 4, all target genes/QTLs (*Sub1A-C, SSIIa, xa5, Xa21, TPS,* qBph3, qBL1 and qBL11) were recombined by crossing the two pseudo-backcrossed BC_2F_1 lines to generate 2,630 pseudo-backcrossed BC_3F_1progeny. Using target MAS and plant-type selection, 158 fully heterozygous pseudo-backcrossed BC_3F_1 lines were selected and selfed (cycle 5) to generate 11,405 F_2 progeny for large-scale, full-target MAS to generate 29 families for the target MAS purification. The numbers of positive plants (pseudo BC_3F_2) for all target genes/QTLs were segregated with Mendelian pattern (homozygous preference genotype = $1/4^n$). In cycle 6, selfing and full-target MAS yielded 29 best-selected, fully homozygous pseudo-backcrossed inbred lines (pseudo-BILs) carrying positive homozygous alleles of all of the donor genes, including *Sub1A-C, SSIIa, xa5, Xa21,* and *TPS* as well as qBph3 and qBL1-qBL11.

Table 1: The four donors and recurrent parents used in the multiple backcross gene pyramiding

Cultivar/ breeding line	Description	Variety type	Cross	Genotype on carrier chromosomes and target gene (s) or QTLs											**References**
				Sub	***xa5***	***Xa21***	**qBL1**	**qBL11**	**Bph3**	***TPS***	***Wx***	SSiia	Os2AP	Hd1	
PinK3	High yielding aromatic rice (pseudo-recurrent parent)	RIL	IR71501/ (KDMl105/ CT9993)	-	-	-	-	-	+	-	+	-	+	+	Rice Science Center (unpublished)
CholSub1	Submergence tolerance aromatic rice with high GT	RIL	IR57514/ KDMl105	+	-	-	-	-	-	-	-	+	+	+	Jantaboon et al. [2011]
Xa497	BLB resistance aromatic rice	BC_2F_2	IR62266/ KDMl105	-	+	+	-	-	-	-	-	-	+	+	Korinsak[2009]
RBPiQ	Blast resistance	RIL	JHN/KDMl105	-	-	-	+	+	-	-	-	-	-	+	Rice Science Center (unpublished)
Bph162	Bph resistance photosensitive rice	BC_3F_6	Rathu Heenati/ KDMl105	-	-	-	-	-	+	+	-	-	-	-	Jairin et al.[2009]

+ = desirable allele.

- = undesirable allele.

Ruengphayak *et al.*

Ruengphayak *et al.* *Rice* 2015 8:7, doi:10.1186/s12284-014-0035-0.

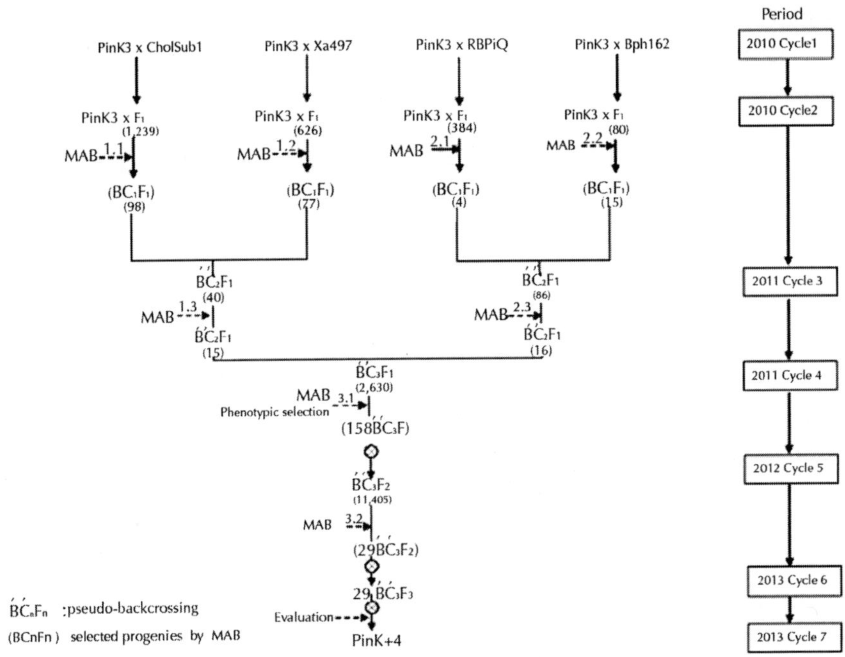

Figure 1: The gene/QTL pyramiding scheme used to generate the high-yield pseudo-BC BIL'PinK+4' line exhibiting submergence tolerance (*Sub1A-C*), bacterial leaf blight resistance (*Xa21,xa5*),rice blast resistance (*qBL1*and*qBL11*), brown planthopper resistance (*qBph3*and*TPS*) and desired cooking qualities (*Wx, SSIIa, Os2AP*).

Of these lines, five pseudo-BC_3F_3 BILs were chosen for field evaluation. The five pseudo BC_3F_3BILs (PinK+4) were selected based on completion of the target genomic regions with interesting starch profiles suitable for glycemic index research in the future (data not shown). In total, four donors and one recipient were intensively crossed and selected for seven cycles: two cycles to generate BC_1F_1, another two cycles to generate pseudo-backcrossed BC_3F_1, and three cycles of selfing to fix the final best-selected pseudo-BC_3F_3BIL for field evaluation.

Graphical Genotyping of Pseudo-BIL

To determine the effects of pseudo-backcrossing on genomic background composition, nine elite pseudo-BIL (BC_3F_2) families were selected for genome scanning using 61 SSRs, 35 and 26 of which were located on six carrier and six non-carrier chromosomes, respectively (Additional file1). The whole genome composition of the selected pseudo-BILs was characterized as the percentages of recurrent genome content (%RGC) and donor genome content (%DGC) based on the physical intervals of the SSR-based genome scanning using five BC_1F_2 (Table 2 and Additional file 2), two BC_2F_2 (data not shown) and nine pseudo-BC_3F_3BILs (Table 3 and Additional file 3). The %RGC and %DGC of BC_1F_2 ranged from 69.99 to 88.98% and from 11.02 to 30.01%, respectively (Additional file 2), whereas those variables for the pseudo-BC_3F_3BILs ranged from 74.50 to 81.30% and 18.70 to 25.50%, respectively (Additional file 3). The mean %RGC and %DGC values for BC_1F_2 were 80.04% and 19.96%, respectively (Table 2), whereas these values for the pseudo-BC_3F_3BILs were 77.48% and 22.52%, respectively (Table 3). These results indicated that at the BC_1F_1 step, there were no significant gains or losses of %RGC and %DGC from the two cycles of pseudo-backcrossing. For this reason, the theoretical RGC in pseudo-BC_3F_3 was not met.

Table 2. The percentage of genome composition (average per target locus) of the selected BC1F2lines resulting from back-crossing of four donors: CholSub1, Xa497, RBPiQ, and Bph162 on PinK3

S	% genome compositions (average per donor)								
	CholSub	Xa497		RBPiQ		qBph162		Ave.	Sum
	Sub1	***xa5***	***Xa21***	**qBL1**	**qBL11**	***TPS***	**qBph3**		
Target gene/QTL	<0.01	<0.01	<0.01	0.03	0.26	<0.01	0.05	0.09	0.34
Donor segment link	1.81	0.00	3.77	1.59	0.00	0.00	0.00	1.79	7.17
Heterozygous segment link[1/]	0.00	0.83	0.13	0.12	0.95	0.00	1.16	0.80	3.19
Donor segment unlink	0.00	0.00	1.03	0.00	0.00	2.60	0.00	0.91	3.63
Heterozygous segment unlink[1/]	0.00	0.00	0.52	0.40	1.61	0.00	0.00	0.63	2.53
Sum of donor segments on target carrier chromosome	1.81	0.83	5.45	2.14	2.82	2.6	1.21	4.22	16.86
Donor segments on non-target carrier chromosome (4-5ch)								7.71	
[1/]Heterozygous segments on non-target carrier chromosome (4-5ch)								1.96	
Sum of donor segments on non-target carrier chromosome								13.89	
Donor segments on non-carrier chromosome (6 ch)								4.29	
[1/]Heterozygous segments on non-carrier chromosome								1.78	
Sum of donor segments on non-carrier chromosome								6.07	
Recurrent background								74.87	
%RGC								80.04	
%DGC								19.96	

[1/]= (1/2 total percentage of physical position).

Ruengphayak *et al.*

Ruengphayak *et al. Rice* 2015 8:7, doi:10.1186/s12284-014-0035-0.

Table 3: The percentage genome compositions (average per target locus) in nine selected pseudo-BC3F3BILs

Region	% genome compositions (average per target)						
	Sub	***xa5***	**qBL1**	***Xa21*_qBL11[1/]**	***TPS***	**qBph3_*SSIIa*[2/]**	**Sum**
Target gene/ QTL	<0.01	<0.01	0.03	0.26	<0.01	0.05	0.34
Donor segment link	1.61	1.22	1.55	2.11	0.19	0.00	6.68
Heterozygous segment link	0.54	0.83	0.24	0.24	0.00	0.00	1.85
Donor segment unlink	0.00	0.18	1.39	1.40	1.44	0.47	4.88
Heterozygous segment unlink	0.00	0.00	0.04	0.23	0.00	0.76	1.03
Sum of donor segments on target carrier chromosome	2.15	2.23	3.25	4.24	1.63	1.28	14.78
Donor segments on non-carrier chromosome (6ch)							5.03
Heterozygous segments on non-carrier chromosome (6ch)							2.71
Sum of donor segments on non-carrier chromosome							7.74
Recurrent background							71.89
% RGC							**77.48**
% DGC							**22.52**

[1/]*Xa21* located within qBL11 on chromosome 11.

[2/]*SSIIa* located within qBph3 on chromosome 6.

Ruengphayak *et al.*

Ruengphayak *et al. Rice* 2015 8:7, doi:10.1186/s12284-014-0035-0.

We then looked into the distribution of donor genome segments across the six carrier and non-carrier chromosomes. For the carrier chromosomes, linkage drags were identified upstream and/or downstream of the donated target gene/QTL following transmission to the recipient genome. Even considering the stringent MAB on all target genes/QTLs during the BC_1 cycle, linkage drags were still detected for almost every donated locus, both homozygous and heterozygous, on one or both sides of the target gene/QTL, constituting more than half of the total donor segments on the carrier chromosomes (Table 2). The largest linkage drags were detected on both sides of the *Xa21* locus (Table 2). However, only a small heterozygous linkage drag was detected at the *xa5* locus. The contrast between the degree of linkage drag for the two functional genes *Xa21* and *xa5* was unexpected, as the pair of loci was inherited from the same donor, Xa497. As both *xa5* and *Xa21* were MAB using their functional markers, it is interesting to speculate on the differences in linkage drag between the two functional genes. The size of the linkage of *Xa21* is much greater than for the two functional genes *Sub1* and *TPS*, and the other QTLs, qBL1, qBL11 and qBph3. Heterozygous linkage drags were identified for five of the eight target genes/QTLs, including the *Xa21* locus (Table 2), Donor-unlinks—the additional donor segments co-transmitted on the opposite (unlinked) chromosome arms of the target genes/QTLs—were identified on the carrier chromosomes containing *Xa21*, qBL1, qBL11 and *TPS* (Table 2). Therefore, the total donor segments transmitted via BC_1 along with *Xa21* were obviously the largest among the target loci under MAB (Table 2).

Multiple target loci from donors are combined in the recipient genome, which include both linkage and non-linkage drags on the carrier chromosomes. Following two successive cycles of marker-assisted pseudo-backcrossing, BC_3F_3BILs linkage drags were detected innearly every case, with the exception of the qBph3-*SSIIa* locus. The total linkage drag (combined homozygous and heterozygous types) was 8.53%, which is more than half of the total donor component on carrier chromosomes (Table 3). The unlinkage drag values of the donated segments on the carrier chromosomes were between 5.91 and 40% of the total donor component on the carrier chromosomes (Table 3). To trace the potential origins of these large donor components, the sum of all donor compositions in the selected BC_1F_2 lines was compared with those components in the selected BC_3F_3BILs. On average, the donor genome content in BC_3F_3BILs was not significantly different from that of BC_1F_2 (Table 2 and Table 3). The same was true for the homozygous components of linkage drag and donor unlinked, whereas the heterozygous components were significantly decreased during pseudo-backcrossing. These results could indicate that high frequency of recombination between donated segments and the recipient genome contents of the carrier chromosomes in the BC_1 cycle was primarily due to the high donor genome content in successive pseudo-backcrossings. Furthermore, the significant reduction in heterozygous donor components for the BC_3F_3BILs highlights an advantage of pseudo-backcrossing in gene pyramiding.

For the non-carrier chromosomes, the total donor components on all six non-carrier chromosomes were not significantly affected (Table 2 and Table 3). However, the heterozygous donor components on both carrier and non-carrier chromosomes persisted, even in the selected pseudo-backcrossed BC_3F_3BILs. These results show that pseudo-backcrossing has similar effects as conventional backcrossing when considering the recovery of the genome content on non-carrier chromosomes.

Evaluation of the Pink + 4 Phenotype

Field evaluations of agronomic and grain quality were performed based on complete target genes/QTLs, plant types, days to harvest and grain quality. The foreground selection successfully fixed homozygosity of the five target genes (*Sub1A-C, SSIIa, xa5, Xa21* and *TPS*) and three QTLs for BL and BPH into approximately 77% of the genetic background of the pseudo-recurrent parent 'PinK3' (Table 3 and Additional file 3). However, the majority of the advanced progeny exhibited significant phenotypic variation from their pseudo-recurrent parent 'PinK3' with respect to nearly all evaluated traits, with the exception of amylose content, grain length per width ratio and polished grain length (Tables 4 and 5).

Table 4: Trait evaluation and agronomic characteristics of four selected pseudo-BC3F3BILs (PinK+4) and parental lines with respect to submergence, bacterial blight, brown planthopper and blast resistance

No	Name	Family	Trait evaluation[1/]				Agronomic characteristics[2/]						
			Sub	BLB (TXO156)	Bph(UBN)	Blast (Mixed#2)	DM	NTP	PH (cm)	**NGP**	**PSF (%)**	**TGW (g)**	**GY (kg/ha)**
1	1E_06	1_H06	50.7$^{c3/}$	1.5^{a}	1.7^{a}	0.7^{a}	140.0^{e}	9.3^{ab}	70.1^{bc}	276.7^{e}	82.5^{ef}	42.9^{f}	7777^{ef}
2	20A09	4_E02	42.0^{bc}	1.4^{a}	1.7^{a}	0.7^{a}	140.0^{e}	9.3^{ab}	79.0^{c}	224.3^{d}	69.6^{c}	41.4^{e}	7381^{ef}
3	66B09	3_E03	31.3^{b}	0.8^{a}	1.7^{a}	0.6^{a}	128.0^{c}	9.2^{ab}	77.7^{c}	236.1^{d}	76.1^{d}	42.5^{f}	6977^{de}
4	78A03	13_H06	52.0^{c}	0.8^{a}	3.7^{abc}	0.0^{a}	143.0^{h}	9.2^{ab}	78.3^{c}	241.0^{d}	77.3^{d}	42.4^{f}	6966^{de}
5	117A08	2_A10	56.0^{c}	1.5^{a}	3.0^{ab}	0.0^{a}	126.0^{c}	8.6^{a}	77.3^{c}	168.7^{c}	81.1^{e}	41.5^{e}	6175^{cd}
6	PinK3	recurrent	0.0^{a}	12.8^{b}	6.3^{cd}	5.7^{d}	141.3^{f}	8.3^{a}	77.0^{c}	301.0^{f}	62.0^{b}	40.3^{d}	8306^{f}
7	Xa497	donor	3.3^{a}	0.5^{a}	6.3^{cd}	3.7^{c}	134.0^{d}	9.5^{ab}	59.3^{a}	166.0^{c}	84.9^{f}	40.5^{d}	5417^{c}
8	CholSub1	donor	50.7^{c}	nd	5.7^{bc}	2.3^{b}	122.0^{b}	10.7^{bc}	62.7^{ab}	120.3^{b}	75.0^{d}	39.4^{c}	4041^{b}
9	RBPiQ	donor	2.0^{a}	18.4^{b}	9.0^{d}	0.0^{a}	142.0^{g}	10.4^{cd}	74.3^{c}	185.7^{c}	94.1^{g}	34.8^{a}	5581^{c}
10	Bph162	donor	0.0^{a}	15.6^{b}	1.0^{a}	3.3^{c}	99.0^{a}	12.8^{d}	69.7^{bc}	81.0^{a}	58.5^{a}	38.5^{b}	2227^{a}

Remarks:

[1/]Trait evaluation.

Sub = Average % plant survival (%PS) after 15 days of flash flooding.

BLB = average lesion length in centimeters of the damage caused by the BB isolate TXO156.

Bph = Severity scores with UBN biotype at 9 DAI when TN1, the susceptible control died.

Blast = Average blast injury score when attacked by a Mixed#2 blast isolate from Thailand.

[2/]Agronomic characteristics.

DM (days to maturity), NTP (number of tillers per plant), PH (plant height from the soil surface to the neck of the panicle), NGP (number of grains per panicle), PSF (percent spikelet fertility), TGW (1,000 grain weight) and GY (grain yield).

3/Average values marked with different letters in the same column are significantly different at the 95% confidence level using LSD.

Ruengphayak *et al.*

Ruengphayak *et al.* *Rice* 2015 8:7, doi:10.1186/s12284-014-0035-0

Table 5: Grain quality and cooking quality traits of five pseudo-BC3F3BILs (PinK+4) as well as the donor and recurrent parents

No.	Name	BR1/(%)	HR1/(%)	GL/W1/	PRL1/(mm)	AC1/(%)	ASV1/	CE1/(%)
1	1E_06	72.9c2	51.8c	3.0ns	0.75ns	29.3c	7.0c	36.7abc
2	20A09	72.9c	52.7c	3.1ns	0.74ns	27.6c	7.0c	43.3cdef
3	66B09	71.5b	52.1c	3.1ns	0.76ns	27.5c	2.0a	31.5a
4	78A03	71.2b	51.3c	3.2ns	0.78ns	27.1c	2.0a	31.2a
5	117A08	73.0c	53.1c	2.9ns	0.73ns	29.5c	7.0c	46.7ef
6	PinK3	72.4bc	43.4b	3.1ns	0.74ns	29.2c	7.0c	39.4bcd
7	Xa497	71.9bc	53.3c	3.4ns	0.74ns	22.0b	5.0b	40.3bcde
8	CholSub	73.2c	42.4b	3.4ns	0.74ns	14.5a	2.0a	44.4def
9	RBPiQ	71.0b	41.3b	3.4ns	0.68ns	14.2a	5.0b	35.9ab
10	Bph162	68.6a	30.2a	3.2ns	0.70ns	14.0a	5.0b	48.6f

Remarks: 1/BR (brown rice), HR (head rice), GL/W (grain length-width ratio), PRL (polished rice length), AC (amylose content), ASV (alkaline spreading value (1.7% KOH)), and CE (cooking elongation).

2/Average values marked with different letters in the same column are significantly different at the 95% confidence level using LSD.

Rice grains were harvested from the yield trial field during the wet season of 2012/2013.

Ruengphayak *et al.*

Ruengphayak *et al. Rice* 2015 8:7, doi:10.1186/s12284-014-0035-0.

4The uniformity of the pseudo-BC_3F_3BILs was the result of MAS. Even with respect to complex traits, such as grain yield, some of these progeny performed as well as the pseudo-recurrent parent 'PinK3' and significantly outperformed their donors (Table 4 and Figure 2). However, some progeny inherited inferior characteristics from the resistance donors, which affected maturity, grain numbers per panicle (NGP), % seed fertility (PSF) and grain yield (GY). In all cases, early-maturing progeny produced a lower grain number per panicle and lower grain yield (Table 4). Therefore, without rigorous background selection, pseudo-backcrossed progeny may not possess the desirable characteristics of their pseudo-recurrent parent.

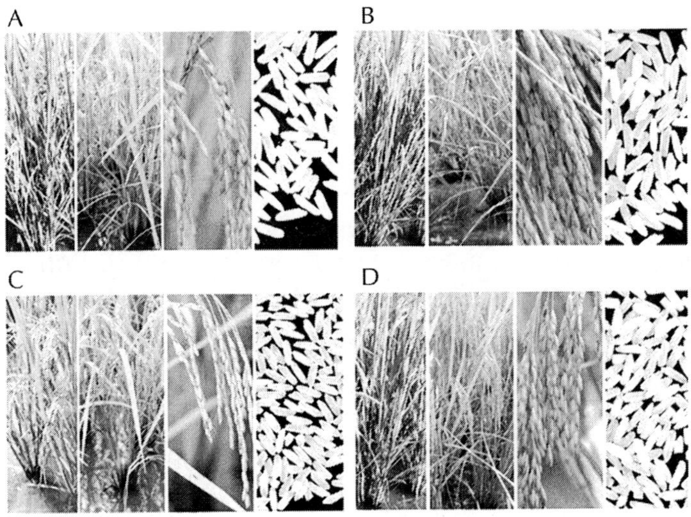

Figure 2: Plant and grain types of pseudo-BC BIL PinK+4 compared with PinK3 (pseudo-recurrent parent). A) PinK+4#1E06, B) PinK+4#20A09, C) PinK+4#66B09 and D) PinK3.

For the *Sub1* selection, all progeny exhibited a significantly improved ability to withstand flash flooding during seedling stages compared with the susceptible 'PinK3' (Table 4). However, significant phenotypic variation (31–56%) was observed among the Sub1-Pink3 families compared with the control phenotype. Two reasons for this variation; First, the *Sub1* gene was not directly inherited from the original FR13A but from CholSub1 donor one of the RILs from the mapping pop 'KD × FR13A' (IR57514) for submergence tolerance QTL by Siangliw et al. ([2003]). The second reason is the quantitative nature of multiple gene/QTL may partially regulated by recipient genetic background of CholSub1 donor and the 'PinK3', the highly susceptible recipient. Furthermore, the fact that significant variations in traits related to submergence tolerance were observed among the individual progeny of IR57514/Kao Dawk Mali 105 (Jantaboon et al. [2011]) is also consistent with our findings, indicating the quantitative nature of such traits.

For BB, selection based on *xa5* and *Xa21* resulted in significant improvements in bacterial leaf blight resistance, specifically to the TXO156 virulent isolate (Table 4) and to several BB isolates identified in Thailand (Additional file 4). For leaf/neck blast resistance, QTL pyramiding of the two QTLs on chromosomes 1 and 11 onto the 'PinK3' background successfully improved resistance to a wide range of blast isolates collected in Thailand (Table 4).

For BPH resistance, based on the UBN biotype (a BPH biotype that has been well characterized in Thailand), the progeny exhibited significant improvement over their pseudo-recurrent parent 'PinK3' but were not as resistant as the resistance donor. The Bph3 QTL and the *TPS* gene from the Bph162 donor were co-inherited from the broad spectrum BPH resistance cultivar 'Rathu Heenati' in crosses with KDML105, as reported by Jairin et al. ([2009]). Some selected introgression lines from that report were moderately resistant (MR) to various BPH biotypes. It appears that introgression of only qBph3 and *TPS* from Bph162 in pseudo-backcrossed BILs was not enough to withstand some of the BPH biotypes used in our experiments.

All selected pseudo-BC_3F_3 BILs (PinK+4) contained the aromatic allele and the *Wxa* allele known to confer high amylose content and a starch profile suitable for further analysis and industrial food applications.

DISCUSSION

Streamline Backcrossing Design

In a gene-pyramiding project, several donors, each providing a target QTL with flanking markers, are used as parental sources for new traits with the goal of improving a favorable variety that has a preferred genetic background. To best facilitate the efficient integration of multiple QTLs into a single optimal variety, an improved breeding platform was developed based on pseudo-backcrossing. In conventional backcross breeding experiments, nearly isogenic lines from each donor were developed prior to pyramiding (Luo and Yin, [2013]; Singh et al. [2013]) to recover the background genome of the recurrent parent. However, when introducing multiple traits, this approach can be tedious and time-consuming. This novel platform based on pseudo-backcross breeding involves both multiple foreground selection and background genome recovery in an abridged manner. Within seven cycles—consisting of a single backcross, two cycles of pseudo-backcrossing and three cycles of line fixation—the entire project was accomplished within four years.

Recurrent Genome Background Recovery

Transferring multiple resistance genes using conventional MAB requires at least three to four backcrosses to guarantee a high recovery of the recurrent parent phenotype (Joshi and Nayak, [2010]; Suh et al. [2013]). In this study, without background selection, the pseudo-BC_3F_3BILs (PinK+4) contained 74.50 to 81.30%RGC, which is significantly below the theoretical value of 93.75% possible following three conventional backcrossings. When comparing graphical genotyping among BC_1F_2 progeny and pseudo-BC_3F_3BILs, the total %DGC was only slightly increased. These results indicate that without background selection, pseudo-backcrossing can only maintain the %RGC gained during the first backcross generation. A low background recovery rate was also reported for the introgression of stripe rust resistance in wheat. Without marker-assisted background selection, the %RGC was only 82% in BC_4F_7 progeny (Randhawa et al. [2009]). However, when combined with phenotypic selection, %RGC was improved to

85–92% in BC_3(Sundaram et al. [2008]; Korinsak et al. [2011]; Singh et al. [2013]). Indeed, the utilization of genome-wide molecular markers for background screening during backcrossing has been suggested as the best method for improving low %RGC (Rajpurohit et al. [2011]; Suh et al.[2013]).

The number of molecular markers used for genome-wide scanning, which ultimately determines cost vs. precision, has varied from 44 to 205 SSR loci in rice. Four groups have reported rice MAB projects involving background selection: Group 1) used 44–51 SSRs (Yi et al. [2009]; Siangliw et al. [2003]; Tomita [2009]; Wongsaprom et al. [2010]), Group 2 used 67–72 SSRs (Win et al.[2012]; Singh et al. [2013]), Group 3 used 84–97 SSRs (Siangliw et al. [2007]; Jantaboon et al.[2011]), and Group 4 used 107–205 SSRs (Rajpurohit et al. [2011]; Korinsak et al. [2011]; Suh et al. [2013]). However, the majority of these MAB projects utilized molecular markers during the final stage of selection.

Linkage Drag

In conventional backcrossing, many portions of the donor genomes are inserted into both the carrier and non-carrier recipient chromosomes during early cycles. After continued backcrossing, the donor genome segments are gradually replaced by sequences from the recurrent parent at varying rates. In most backcrossing programs, linkage drag is responsible for long-lasting donor genome segments remaining in the recurrent genome. However, there is no difference in terms of linkage drag in pseudo-backcrossing schemes. In pseudo-backcrossing BILs, there was only a 1% reduction in linkage drag from BC_1 to pseudo-BC_3, and most of this reduction was due to the heterozygous segments of the linkage drag. Our results also revealed that the degree of linkage drag is independent of the size of the target genes/QTLs. When comparisons were made between the selection of single genes or single QTLs, the degree of linkage drag was less for QTLs than for single genes. Of the five single gene selections, *SSIIa, Sub1, xa5, Xa21* and *TPS*, only selection for *Xa21* showed large, persistent linkage drag. Persistent linkage drags when selecting for BB genes have been reported in several backcross breeding programs, such as during the transfer of*Xa4+xa5+Xa21* from indica 'IRBB57' into japonica 'Mangeumbyeo' (Suh et al. [2013]), and pyramiding of the BB resistance genes *Xa21* and *xa13* and a semi-dwarfing gene (*sd-1*) from PR106-P2 into Type 3 Basmati (Rajpurohit et al. [2011]). The problem of persistent linkage drag when selecting for *Xa21* may due to the fact that *Xa21* was derived from IRBB21, which inherited the chromosomal region containing *Xa21* from the wild species *Oryza longistaminata* through several cycles of backcrossing with indica rice (Song et al. [1995]). The degree of linkage drag may depend on linkage disequilibrium surrounding the target gene to be transferred. Genes inherited from wild species in cultivated strains may be retained within long, stable LD stretches that are difficult recombine. Therefore, selection based on functional markers alone does not guarantee linkage drag-free progeny. More successful single-gene target selection has been reported when markers flanking the gene of interest were also selected for (Rajpurohit et al.[2011]). In one of the most comprehensive backcrossing projects, a single gene, *sd1,* was integrated into the desirable variety 'Koshihikari' using eight cycles of MAB with 51 SSR markers surrounding *sd1* to completely eliminate linkage drag (Tomita [2009]).

Grain Yield Performance of Pseudo-BC BILS

In this study, we combined multiple resistance genes from four donors using a new backcrossing method involving pseudo-backcrossing. The results show that all pseudo-BC BILs showed significant improvements in resistance to BB, BL, BPH and Sub compared with the pseudo-recurrent parent 'PinK3', as well as significant improvements in grain yield (21–68% over the donors, but 7–26% lower than the recipient). The reduction in grain yield in the pseudo-BC BILs should be interpreted in several ways. First, there was an average of 7.6% and 22.5% of linkage drag and DGC, respectively, in the recurrent genome. As these donors were inferior in grain yield with respect to the pseudo-recurrent parent 'PinK3', the high persistent %DGC could have disrupted the optimal expression of high-yield genes in pseudo-BC BILs. Second, there can be a slow recovery of the %RGC during pseudo-backcrossing when MAS for the recurrent background is not in place. Under such conditions, reconstruction of the recurrent genome content by recombining different pseudo-backcrossed lines is not favorable, as different donor segments on both carrier and non-carrier chromosomes have more chances to recombine, creating new substitution lines that do not resemble the recurrent parent. Third, the multiple donated genes/QTLs from donors to the recipient act as a 'genetic load' against the fitness of the recurrent parent. The over-expression of multiple resistance genes could counteract the metabolic energy needs necessary for high yield. Therefore, pseudo-backcrossing may be the fastest method for gene/QTL pyramiding, although it may not be the ideal breeding platform for creating elite recurrent varieties. However, marker-assisted, genome-wide scanning can be implemented during early stages to facilitate the reconstruction of favorable genomic backgrounds at the end of the pseudo-backcrossing scheme. Ultimately, the trade-offs must be considered by the breeders. Ideally, new high-throughput, low-cost, genome-wide scanning technologies should be utilized in combination with skillful breeder selection. For the whole project, more than 50,000 plants were individually genotyped for at least one molecular marker. If budget is allow, extensive background selection must be emphasized.

CONCLUSIONS

We improved high-yield, aromatic rice varieties by introducing desirable multiple traits by pyramiding five target genes (*Sub1A-C, SSIIa, xa5, Xa21* and *TPS*) and three QTLs (qBph3, qBL1 and qBL11) from four resistance donors. We redesigned the gene-pyramiding platform to minimize the total project time span by integrating MAS into pseudo-backcross breeding. Consequently, only seven breeding cycles in four years were required to develop new varieties exhibiting multiple resistance traits. Using pseudo-backcrossing, approximately 77.48% of the recurrent genome background was recovered. With additional background genome selection, the recurrent genome background can further improve the %RGC and optimize the expression of introgressed QTLs.

METHODS

Plant Materials Used

PinK3 is a high-yield, irrigated aromatic rice cultivar developed by Rice Science Center, Kasetsart University, Thailand (unpublished).

However, this variety is susceptible to flash flooding (Sub), bacterial leaf blight (BB), leaf-neck blast (BL) and the brown planthopper (BPH). The four donors used to transfer five genes and three QTLs to the pseudo-recurrent parent 'PinK3' (Table 1) were developed by the Rice Gene Discovery and Rice Science Center, Kasetsart University, Thailand. The four donors used to improve the abiotic and biotic stress tolerances of 'PinK3' are listed in Table 1.

Pseudo-Backcross Design

The pseudo-backcross platform is divided into three steps. In the initial step, one round of backcrossing is conducted to donate the favorable QTL allele to the recipient background using marker-assisted backcrossing (MAB). Each QTL-BC_1F_1 contains approximately 75% of its recurrent genome content (RGC). In the second step, the BC_1F_1-plus QTLs are used as pseudo-recurrent parents, and pseudo-BC_2F_1 plants are formulated by crossing between them. More BC_1F_1 plus new QTLs can be crossed to successively generate pseudo-BC_nF_1 QTLs and, thus, to continue streamline gene pyramiding. At the end of this step, the BC_nF_1, which contains the new QTLs at full heterozygosity at all target marker loci, are self-pollinated to fix the target loci (Additional file 5). In rice, this new platform allows breeders to stack more QTLs in the shortest possible time (shorter than that required by any other method).

Donors for Gene Pyramiding in Rice

Four donors providing submergence tolerance, bacterial leaf blight resistance, blast resistance, BPH resistance and desired gelatinization temperature were introduced into PinK3 (*aroaro* and *WxAWxA*) as the female pseudo-recurrent parent. The donor CholSub1 carries two target traits: submergence tolerance and desired gelatinization temperature (*Sub1A-C* and *SSIIa*); the donor Xa497 carries two functional genes for bacterial leaf blight resistance (*xa5* and *Xa21*); the donor RBPiQ carries two QTLs for blast resistance (qBL1 and qBL11); and the donor Bph162 carries two target traits for BPH resistance (qBph3 and *TPS*) (Table 1). These four donors were used for gene pyramiding.

Genomic DNA Isolation

Rice seedlings from each segregating population were grown in a 288-well plastic tray (representing three 96-well plates). Young leaves from 14-day-old individual plants were cut into small pieces and placed (~0.2 g weight per sample) in a 2-ml 96-well plastic block. Leaf tissues were ground in liquid N_2 using a Tissue Striker II (KisanBio, Seoul, South Korea). After grinding, 300 µl Agencourt®Chloropure lysis buffer was added to the samples. Homogenized tissues were incubated in a 2-ml 96-well plastic block at 65°C for 1 hour. The sample blocks were then centrifuged at 4,000 rpm for 10 minutes. Lysates (containing at least 200 µl) were transferred to a new 2-ml 96-well plastic block using an Automated Biomeck NX AP96 instrument (Beckman Coulter, California, USA). The extraction was conducted using the standard protocol of Agencourt Chloropure for nucleic acid isolation from plants (Beckman Coulter, California, USA).

Foreground MAS

Foreground selection was performed using two marker systems. For SNP and functional markers, multiplex genotyping was conducted using the SNPstream system (Beckman Coulter, California, USA). The remainder of the foreground markers were SSR markers flanking specific QTLs. The SNP-based genotyping array was described.

High-Throughput Multiplex SNP Genotyping

High-throughput genotyping was performed by multiplex PCR, as described by Bell et al. ([2002]) with certain modifications. In brief, the forward/reverse (18–20 nt in length) and SNP-specific (40–45 nt in length) primers were designed for each foreground locus (Primers can be manually designed) to generate a product of 90–180 nt in size. The program selects the best Single Base Extension (SBE)-primer based on sequence melting temperature (T_m; °C) and secondary structure. At the 5′ end of the SBE-primer sequences are 20-nt tags that are complementary to the sequences of specific positional tags in the SNPware (384-well) microarray format (Beckman Coulter, California, USA) (Table 6).

Table 6: Primer sequences used in the SNP genotyping format for foreground selection

No.	SNP name	Chr.	Trait	Tag number	SNP type	Amplicon size (bp)	Forward primer (5′---3′)	Reverse primer (5′---3′)	SNP-specific primer (5′---3′)
1	Sub1C	9	Sub	9	A/G	111	ACGAGCCGACG-ACGACGA	ATCTCCGACGC-CCACCTC	CCGCCAGTAAGACCTAGACGCGGC-GGCGGCGGCGGCGGAGGGAGA
2	xa5*	5	BB	2	C/G	303	GGCCACCTTCG-AGCTCTACC	CAACATTGCAAC-TCCGTGATAAG	CTCAGACTACGAATCCACGTGTAAAGTA-GATACCTTATCAAACTG
3	Xa21	11	BLQ11	40	T/C	141	AAAGCTAGGCT-GCTAGTGCTG	AAAATAGTATATAT-GTACCACTGCTTCTT	AAGTACCACGTCAACGTCACTATGCTTC-AAGGTCAGGGTGGTCGA
4	SSIIa	6	GT	25	G/T	89	CCACTGCCTCG-AGACGTA	CGTGGTCCCAG-CTGAGGT	CCATAACAACTTACCAGCCAGCAAGTAC-AAGGAGAGCTGGAGGGG
5	Waxy	6	AC	31	G/T	141	TTCACTTCTCTG-CTTGTGTTGT	TACTTGTAAGGA-AAAACGAGCAA	CAGAACATCCTCAGAAGCAAGTTCATCA-GGAAGAACATCTGCAAG
6	*aromarker**	8	FR	41	A/G	400	AATCATGTATAC-CCCATCAA	TTTCCACCAAGT-TCCAGTGA	TACCTATGACCAGCAAGCACAACCTTAACCA-TAGGAGCAGCTGAA
7	Hd1	6	Photo	13	C/G	152	TCCAAAGATTCC-GACAACA	TTGTCGTAGTAC-GAATTGTACCC	CAACAATACGAGCCAGCAAGACAACAA-CAACGACAACGACAATAA
8	TPS_Chr4_ATHB1*	4	Bph3	27	T/G	264	AAGCGCTTATATT-CAAGCAGAA	TCCATTCTTCCG-ATCTCTGG	GCAAGCCATCAGCTAATACATTCAT-GAAACAGTTCTAGCAATAAT
9	Bph_Chr6_1210*	6	Bph3	17	T/C	287	GAAAGCCTTTGAA-ACAAAGTATTGA	CTTGAATTTGAA-GTTGATTTTAGGG	GCAGACAACGAACAACTACCAAA-CGGCATATTTGCAAACAGAAAA
10	Bph_Chr6_3380*	6	Bph3	22	T/C	300	AGAGGAAATGAT-TCAAGGAG	AGCTAGCAGGC-GTAGCTTAT	GCAACATAAGACCGCTCAACTAG-TTAATTTCACGCCATGACAGAT
11	Bph_Chr6_1380*	6	Bph3	19	A/G	296	TTTTGTTTTCTTCT-TGAGAGTGGTC	CAAGGTAATGA-CATCAAGAACCAAT	CCACTCAACTCCACGAATACCAC-ATGTTTATTTTTAATTTCACAG
12	TBGI055716	1	BLQ1	24	T/A	169	ACGATGCGGCA-CTCGTCG	TGTTCTTGAAC-GCGGCGA	CAACAAGACATAACAACGCAGT-GGAGTGGTGGATGAAGCGGAAGA
13	TBGI055578	1	BLQ1	42	A/G	93	ATTTGCTGCTC-ATGGTGG	TGGGGAAGCCG-AGGAGAT	CTCACTATCTGACAAGCCACGT-GGAGCCTCCTCACCAGGAAGTGC
14	TBGI055841	1	BLQ1	46	T/C	115	ATTGGCATCGTT-TGGTCTG	ATTCCGTGCATA-TATACGAACTTC	ACAGATCACTCACCGACTAAATAC-CTGCGTCGAGTAGAGACGATG
15	TBGI454069	11	BLQ11	14	A/G	142	TGAATTGTCGTC-CTCTAACAAC	TATGGAATATCT-GCATCATGAAC	AACATACAGACGCACTCCTCTCTTTG-CTATGCCAAAGTCTGCTAC

16	TBGI453598	11	BLQ11	23	C/T	90	TTTGCTGTGAC--GGGAAGA	AAAAAGGAACT-AGCCAGTTTTG	AGTAGCCTAACAGCACTCGAAAGAT-CGAGTGCTCTATTGCAACCG
17	TBGI453126	11	BLQ11	32	C/T	92	ACCGACGCT-GCTGCAGAA	AGCGGTTATGt-GATGGCTAA	CAAGCAACGACCTACTACAAGCATG-GCGTTTGAGCGCGTCCTGGG
18	TBGI454717	11	BLQ11	36	T/A	132	ATCCTACCG-CCGCTCTG	AATTCGGTCTTCt-GTAAACACG-	CACCGCTATCAACAGACTTGGTGCG-GTAGTTTCTGGGAAGCTACG
19	TBGI454800	11	BLQ11	38	G/C	113	TACTACAAC-AACAGGAACGCC	TTGATGATGAAG-TGtGATGAGC	ACGTAAGACCACTCAAGACCAGAAG-ACGCTGAACAGGATGGCGAT

*These primer designs were optimized manually.

Ruengphayak *et al.*

Ruengphayak *et al.* *Rice* 2015 8:7, doi:10.1186/s12284-014-0035-0

Multiplex PCR

A 10-μl PCR reaction containing 5 μl genomic DNA (10 ng/μl) and 5 μl KAPA TaqHotStart PCR buffer (1 U KAPA TaqHotStart (KapaBiosystems, MA, USA) (final concentration of 1× KAPA TaqHotStart Buffer: 75 μM dNTPs, 5 mM $MgCl_2$ and 50 nM 38-primer pool) was performed in a 384-well PCR plate (Sorenson BioScience, UT, USA). The following thermocycler touch-up PCR cycle was used: 95°C for 3 min, followed by 6 cycles of 95°C for 30 seconds, 50°C up to 55°C (0.3 increment/cycle) for 30 seconds, and 72°C for 30 seconds; this was followed by 34 cycles of 95°C for 30 seconds, 55°C for 30 seconds, 72°C for 30 seconds and a final extension at 72°C for 7 min. Subsequently, the temperature was held at 4°C.

Multiplex PCR assays were prepared separately based on the SNP panel type (A/G, A/C, A/T, G/C, G/T and C/T). Following PCR amplification, PCR products were cleaned, and the SBE reactions were performed. Next, multiplex SBE products from different panel types were pooled prior to the hybridization step.

PCR clean up, SBE reactions, hybridization and washing, SNPstream imaging and data analysis were performed as described by Bell et al. ([2002]).

Evaluation of Abiotic and Biotic Stress Traits

Submergence Screening

The parents and pseudo-BC_3F_3BILs (PinK+4) were screened for submergence resistance traits. The experiment was conducted under complete submergence in an outdoor lagoon located at the Rice Science Center, Kasetsart University, Kamphaeng Sean Campus, Thailand, during the dry season of 2013. The experiment was arranged using a randomized complete block design with three replications. Sixteen three-week-old BC_3F_3BIL seedlings and controls including PinK3, CholSub, Xa497, RBPiQ and Bph162 were transplanted in three replicate plots (plot size: 0.75 × 0.75 m^2) at a spacing of 25 cm × 25 cm. Two weeks after transplanting, the number of seedlings was counted in each plot; then, the lagoon was filled with water to a depth of 2 m. To impose severe submergence stress, the seedlings were completely submerged for 15 days; the water level was maintained at 1–1.2 m above the leaf tip throughout the experimental period. After this period, the lagoon was drained, and the seedlings were re-exposed to air for 10 days (Jantaboon et al. [2011]). The number of surviving plants was recorded. The percentage of survival (PS) was calculated using the following equation:

$$PS = \frac{\text{Number of surviving plants}}{\text{Total number of plants}} \times 100$$

Brown Planthopper (BPH) Screening

A set of pseudo-BC_3F_3BILs (PinK+4) and their parents were screened for resistance against BPH using standard seedbox screening (SSBS); the BPH population used was collected from Ubon Ratchathani provinces (Jairin et al. [2007]). The SSBS was conducted at the seedling stage (10 days old) under greenhouse conditions following the method described by Heinrichs et al. ([1985]). Damage scores were recorded when the susceptible control, 'TN1', died (9 days after infestation; 9DAI), using the standard evaluation system (IRRI [1996]).

Bacterial Leaf Blight Screening

The *Xoo* isolate TXO156 was selected for this experiment. The isolate was grown following the method described by Win et al. ([2012]). The parents and pseudo-BC_3F_3BILs (PinK+4) were grown in a greenhouse for 30 days before inoculation. The inoculation procedures used were adapted from those described by Korinsak et al. ([2009a], [b]). Three to four fully expanded leaves of each plant were inoculated. Lesion length (LL) was measured at 12–14 days after inoculation. The resistance reaction was classified as resistant (R), moderately resistant (MR), moderately susceptible (MS) and susceptible (S) when the values of LL were 0–3 cm, 3.1–6 cm, 6.1–9 cm and more than 9 cm, respectively (Yang et al. [2003]; Lin et al. [1996]).

Leaf Blast Screening

Thailand *Magnaporthe oryzae* mixed isolates#2, including THL710 (Mae Hong Son), THL282 (Phrae), THL906 (Yala), THL122 (Chiang Rai), THL757 (Mae Hong Son) and THL603 (Surin) (Rice Gene Discovery, Thailand, unpublished), which can damage the PinK3 form in the mixed isolate pre-screening, was used in leaf blast screening experiments. The inoculum was prepared and the plants were inoculated following the method described by Marchetti et al. ([1987]) with some modifications. Pseudo-BC_3F_3BILs (PinK+4) and their parents were grown in polyvinyl trays containing paddy field soil (four seedlings per line) following a three-replication completely randomized design (CRD). The seedlings were maintained in a greenhouse for 17 days before inoculation, after which they were inoculated with mixed isolates#2. Disease scoring was recorded at seven days after inoculation on a 0 to 6 scale following the procedure described by Roumen et al. ([1997]) and IRRI ([2002]). The average score of each line was computed from the disease score measured for 12 individual plants.

Recording of Important Agronomic Traits

Traits measured included days to 100% flowering (DF100), days to maturity (DM), number of tillers per plant (NTP), plant height (PH), number of grains per panicle (NGP), percent spikelet fertility (PSF), 1,000 grain weight (TGW) and grain yield (GY); these traits were measured for rice plants grown during field trials at Kasetsart University, Kamphaeng Sean, Nakhon Pathom, Thailand.

Twenty-one-day-old seedlings were transplanted in three replicates in 1 × 2 m^2 plots using 25 × 25 cm^2 plant spacing. Agronomic traits were recorded for five randomly selected plants grown in each plot. The DF was recorded when 100% of the individual plants in each plot flowered. NTP, PH, NGP and DM were measured at maturity, and the results were averaged from five randomly selected plants in each plot. PH was measured from the soil surface to the neck of the panicle. The NGP was counted manually for five panicles. To measure the GY in each plot, only the inner rows (containing 21 plants) were used. Two border rows on each side of the plot and the border plants of each row were discarded. The GY recorded for each plot was adjusted to

14% moisture content and then extrapolated to units of kg per ha. TGW measurements were replicated three times. Statistical analysis was performed using the STATGRAPHICS plus 3.0 software package (Manugistics [1997]).

Evaluation of Grain Quality

Grain quality was evaluated using grain harvested from the trials field. Rice grains of the pseudo-BC_3F_3BILs (PinK+4) and their parents were harvested at physiological maturity and sun-dried in a greenhouse. The dried grains were stored at room temperature for one month prior to the grain quality traits evaluation. Three hundred grams of grains was sampled from each replicate. The grains were mechanically dehulled and polished using a mini-polisher. Four physical grain qualities, including percentages of brown rice (BR), head rice (HR), grain length (GL/W) and %cooking elongation increased (CE), were evaluated for the polished rice. Ten grains of paddy rice were measured using a Vernier caliper, and the GL/W ratio was calculated. The polished rice grain length (PRL) was measured using the same method. The cooking elongation of the polished rice was determined by boiling 20 grains in 5 ml of dH_2O for ten minutes. Cooked grain lengths were measured after air-drying the grains for 1 hour. Two chemical grain qualities, amylose content (AC) and gel temperature (GT), were evaluated following the procedures described by Lanceras et al. ([2000]). GT is an indicator of the time required for cooking. The GT was indirectly estimated based on the alkali spreading value (ASV); higher values of ASV represent increased spreading in alkali and therefore represent lower values of GT; conversely, smaller values of ASV indicate higher values of GT.

Genomic Scanning of Foreground and Background

Using simple sequence repeat (SSR) markers spanned across the genome, the effects of foreground selection on linkage drag and genome background content were estimated using the graphical genotyping software GGT 2.0 (Berloo [2007]) based on a specific F_2 population type calculation. The effects of marker-assisted selection were studied on BC_1F_2 and pseudo-backcross progeny after recombining all target genes/QTLs (PinK+4). Five selected BC_1F_2 (Additional file 2) and nine pseudo-BC_3F_3BILs (PinK+4) (Additional file 3) representing nine recombined families were analyzed for genetic background recovery; 61 SSR markers showing clear polymorphism between parents were used (Additional file 1). Twenty-six markers distributed over six non-carrier chromosomes (chromosomes 2, 3, 7, 8, 10 and 12), as well as 35 SSR markers distributed over six carrier chromosomes (chromosomes 1, 4, 5, 6, 9 and 11) were used for background scanning. Furthermore, QTL-specific markers located within each QTL were developed to estimate the risk of target loss. Definitions of parameters describing % recurrent genome content (%RGC) and % donor genome content (%DGC) were calculated according to Xi et al. ([2006]), (Suh et al. ([2009]) and Suh et al. ([2013]) with some modification.

Donor segment link: homozygous allele similar to each donor on the same arm of the carrier chromosome.

Heterozygous segment link: heterozygous allele on the same arm of the carrier chromosome.

Donor segment unlink: homozygous allele similar to each donor on the other arm of the carrier chromosome.

Heterozygous segment unlink: heterozygous allele on the other arm of the carrier chromosome.

Donor segments on non-carrier chromosome: homozygous allele similar to each donor on non-carrier chromosomes.

Heterozygous segments on non-carrier chromosomes: heterozygous allele on non-carrier chromosomes.

Recurrent background: homozygous allele similar to the pseudo-recurrent 'PinK3' allele.

AUTHORS' CONTRIBUTIONS

SR and SP performed the cross-hybridization and MAB screening experiments. EC developed the SNP marker set used for MAB. WK and WS performed the BPH infestation experiments and identified candidate genes for BPH resistance. EP performed the submergence screening. SK and SK performed the BB and BL screenings. The entire study was designed and coordinated by AV. SR drafted the manuscript, and AV corrected the manuscript's draft. All authors read and approved the final version of the manuscript.

ACKNOWLEDGMENTS

This work was supported by the Agriculture Research Development Agency (ARDA) (Grant No. P12/2552). SR gratefully acknowledges financial support from the Royal Golden Jubilee (RGJ)-PhD program, Grant No. PHD/0009/2546, from the Thailand Research Fund (TRF).

REFERENCES

1. Ashikari M, Matsuoka M: Identification, isolation and pyramiding of quantitative trait loci for rice breeding. *Trends Plant Sci* 2006, 11(7):344-350.
2. Bao JS, Sun M, Zhu LH, Corke H: Analysis of quantitative trait locus for some starch properties in rice (*Oryza sativa* L.), thermal properties, gel texture, swelling volume. *J Cereal Sci* 2004, 39:379-385.
3. Bao JS, Corke H, Sun M: Nucleotide diversity in starch synthase IIa and validation of single nucleotide polymorphisms in relation to starch gelatinization temperature and other physicochemical properties in rice (*Oryza sativa* L.). *Theo App Genet* 2006, 113:1171-1183.
4. Bao JS, Shen SQ, Sun M, Corke H: Analysis of genotypic diversity in the starch physicochemical properties of non-waxy rice: apparent amylose content, pasting viscosity and gel texture. *Starch-Starke* 2006, 58:259-267.
5. Bell PA, Chaturvedi S, Gelfand CA, Huang CY, Kochersperger M, Kopla R, Modica F, Pohl M, Varde S, Zhao R, Zhao X, Boyce-Jacino MT: SNPstream UHT: ultra-high throughput SNP genotyping for pharmacogenomics and drug discovery. *Biotechniques* 2002, 32:S70-S77.
6. Berloo RV: *GGT 2.0 user manual. Laboratory of plant breeding.* Wageningen University, Netherland; 2007.
7. Bouman BAM, Lampayan RM, Tuong TP (2007) Water management

in irrigated rice: coping with water scarcity. International Rice Research Institute (IRRI), Los Baños, Philippines. http://www.knowledgebank.irri.org/ewatermgt/courses/course1/resources/books/water_management_1.pdf.

8. Bouquet A: Introduction dansl'espece *Vitis vinifera* L. d'un caractere de resistance al'oidium (*UncinulanecatorSchw.* Burr) issul'espece *Muscadinia rotundifolia* (Michx) Small. *Vignevini* 1986, 12(suppl):141-146.
9. Heinrichs EA (1979) Chemical control of the brown planthopper. In: Brown planthopper: Threat to rice production in Asia, IRRI Publications, Los Baños, Philippines, pp. 145-167.
10. Heinrichs EA, Medrano FG, Rapusas HR: *Genetic Evaluation for Insect Resistance in Rice*. IRRI, Los Ban˜os, Philippines; 1985.
11. IRRI: *Standard Evaluation System for Rice*. IRRI, Manila, Philippines; 1996.
12. IRRI: *Standard Evaluation System for Rice (SES)*. International Rice Research Institute, Los Banos, Philippine; 2002.
13. Jairin J, Phengrat K, Teangdeerith S, Vanavichit A, Toojinda T: Mapping of a broad-spectrum brown planthopper resistance gene, *Bph3*, on rice chromosome 6. *Mol Breed* 2007, 19:35-44.
14. Jairin J, Teangdeerith S, Leelagud P, Kothcharerk J, Sansen K, Yi M, Vanavichit A, Toojinda T: Development of rice introgression lines with brown planthopper resistance and KDML105 grain quality characteristics through marker-assisted selection. *Field Crop Res* 2009, 110(3):263-271.
15. Jantaboon J, Siangliw M, Im-mark S, Jamboonsri W, Vanavichit A, Toojinda T: Ideotype breeding for submergence tolerance and cooking quality by marker-assisted selection in rice. *Field Crop Res* 2011, 123(3):206-213.
16. Jena KK, Jeung JU, Lee JH, Choi HC, Brar DS: High-resolution mapping of a new brown planthopper (BPH) resistance gene, *Bph18*(t), and marker-assisted selection for BPH resistance in rice (*Oryza sativa* L.). *Theor Appl Genet* 2006, 112:288-297.
17. Joshi RK, Nayak S: Gene pyramiding-a broad spectrum technique for developing durable stress resistance in crops. *Biotechnology and Molecular Biology* 2010, 5(3):51-60.

18. Kamolsukyunyong W, Sukhaket W, Ruanjaichon V, Toojinda T, Vanavichit A: Single-feature polymorphism mapping of isogenic rice lines identifies the influence of terpene synthase on brown planthopper feeding preferences. *Rice* 2013, 6:18.

19. Kate-ngam S, Kimchaiyong W, Wanchana S, Toojinda T: Association analysis and functional marker development of soluble starch synthase IIa (SSIIa) and gelatinization properties in Thai rice. In *Proceeding of the 5th International Crop Science Congress & Exhibition: Recognizing Past Achievements, Meeting Future Needs April 13–18, 2008*. International Convention Center, Jeju, Korea; 2008.

20. Kawaguchi M, Murata K, Ishii T, Takumi S, Mori N, Nakamura C: Assignment of a brownplanthopper (*Nilaparvatalugens*Stål) resistance gene *bph4* to the rice chromosome 6. *Breed Sci* 2001, 51:13-18.

21. Korinsak S. Marker-assisted pyramiding bacterial blight resistance genes (xa5, Xa21, xa33(t), Xa34(t) and qBB11) in rice. Dissertation: Kasetsart University; 2009.

22. Korinsak S, Sirithanya P, Toojinda T: Identification of SSR markers linked to a bacterial blight resistance gene in rice cultivar 'Pin Kaset'. *KKU Res J (GS)* 2009, 9:16-21.

23. Korinsak S, Sriprakhon S, Sirithanya P, Jairin J, Korinsak S, Vanavichit A, Toojinda T:Identification of microsatellite markers (SSR) linked to a new bacterial blight resistance gene xa33 (t) in rice cultivar 'Ba7'. *Maejo Int J Sci Technol* 2009, 3:235-247.

24. Korinsak S, Sirithunya P, Meakwatanakarn P, Sarkarung S, Vanavichit A, Toojinda T:Changing allele frequencies associated with specific resistance genes to leaf blast in backcross introgression lines of Khao Dawk Mali 105 developed from a conventional selection program. *Field Crops Res* 2011, 122(1):32-39.

25. Kullan ARK, Van Dyk MM, Jones N, Kanzler A, Bayley A, Myburg AA, Myburg Z: High-density genetic linkage maps with over 2,400 sequence-anchored DArT markers for genetic dissection in an F_2 pseudo-backcross of *Eucalyptus grandis* × *E. urophylla*. *Tree Genetics & Genomes* 2012, 8:163-175.

26. Lanceras J, Huang ZL, Naivikul O, Vanavichit A, Ruanjaichon V, Tragoonrung S: Mapping of genes for cooking and eating qualities in Thai jasmine rice (KDML105). *DNA Res* 2000, 7:93-101.

27. Lin XH, Zhang DP, Xie YF, Gao HP, Zhang Q: Identifying and mapping a new gene for bacterial blight resistance in rice based on RFLP markers. *Phytopathology* 1996, 86:1156-1159.

28. Lu Y, Xiao P, Shao Y, Zhang G, Thanyasiriwat T, Bao J: Development of new markers to genotype the functional SNPs of SSIIa, a gene responsible for gelatinization temperature of rice starch. *J Cereal Sci* 2010, 52:438-443.

29. Luo Y, Yin Z: Marker-assisted breeding of Thai fragrance rice for semi-dwarf phenotype, submergence tolerance and disease resistance to rice blast and bacterial blight. *Mol Breeding* 2013, 32:709-721.

30. Maclean JL, Dawe DC, Hardy B, Hettel GP: *Rice Almanac*. 3rd edition. IRRI, WARDA, CIAT and FAO, Philippines; 2002.

31. Manugistics. Statgraphics Plus for Windows 3.0. Manugistics Group, Inc., Rockville: Maryland, USA; 1997.

32. Marchetti MA, Lai X, Bollich CN: Inheritance of resistance to *Pyriculariaoryzae*in rice cultivar grown in the United States. *Phytopathology* 1987, 77:799-804.

33. Masouleh AK, Waters DLE, Reinke RF, Ward R, Henry RJ: SNP in starch biosynthesis genes associated with nutritional and functional properties of rice. *Sci Rep* 2012, 2:557.

34. Molnár S, Galbács Z, Halász G, Hoffmann S, Kiss E, Kozma P, Veres A, Galli Z, Szőke A, Heszky L: Marker assisted selection (MAS) for powdery mildew resistance in a grapevine hybrid family. *Vitis* 2007, 46(4):212-213.

35. Montoya C, Cochard B, Flori A, Cros D, Lopes R, Cuellar T, Espeout S, Syaputra I, Villeneuve P, Pina M, Ritter E, Leroy T, Billotte N: Genetic architecture of palm oil fatty acid composition in cultivated oil palm (*Elaeis guineensis* Jacq.) Compared to its wild relative *E. oleifera* (H.B.K) Cortés.*PLoS ONE* 2014, 9(5):e95412. doi:10.1371/journal.pone.0095412

36. Noenplab A, Vanavichit A, Toojinda T, Sirithunya P, Tragoonrung S, Sriprakhon S, Vongsaprom C: QTL mapping for leaf and

neck blast resistance in KhaoDawk Mali 105 and Jao Horn Nin recombinant inbred lines. *Sci Asia* 2006, 32(2):133-142.

37. Novaes E, Osorio L, Drost DR, Miles BL, Boaventura-Novaes CR, Benedict C, Dervinis C, Yu Q, Sykes R, Davis M, Martin TA, Peter GF, Kirst M: Quantitative genetic analysis of biomass and wood chemistry of *Populus* under different nitrogen levels. *New Phytol* 2009, 182:878-890.
38. Ou SH:*Kew Surrey: Common wealth Mycological Institute*. 2nd edition. 1985.
39. Rajpurohit D, Kumar R, Kumar M, Paul P, Awasthi AA, Basha PO, Puri A, Jhang T, Singh K, Dhaliwal HS: Pyramiding of two bacterial blight resistance and a semi dwarfing gene in type 3 Basmati using marker-assisted selection. *Euphytica* 2011, 178:111-126.
40. Randhawa HS, Mutti JS, Kidwell K, Morris CF, Chen X, Gill KS: Rapid and targeted introgression of genes into popular wheat cultivars using marker-assisted background selection. *PLoS ONE* 2009, 4(6):e5752.
41. Roumen E, Levy M, Nottegham JL: Characterization of the European pathogen population of *Magnaporthe grisea* by DNA fingerprinting and pathotype analysis. *European J Pl Path* 1997, 103:363-371.
42. Servin B, Martin OC, Mézard M: Toward a theory of marker-assisted gene pyramiding. *Genetics* 2004, 168:513-523.
43. Siangliw M, Toojinda T, Tragoonrung S, Vanavichit A: Thai jasmine rice carrying QTLch9 is submergence tolerance. *Ann Bot* 2003, 91:255-261.
44. Siangliw JL, Jongdee B, Pantuwan G, Toojinda T: Developing KDML105 backcross introgression lines using marker-assisted selection for QTLs associated with drought tolerance in rice. *ScienceAsia* 2007, 33:207-214.
45. Singh A, Singh VK, Singh SP, Pandian RTP, Ellur RK, Singh D, Bhowmick PK, Krishnan SG, Nagarajan M, Vinod KK, Singh UD, Prabhu KV, Sharma TR, Mohapatra T, Singh AK (2012) Molecular breeding for the development of multiple disease resistance in Basmati rice. AoB PLANTS: pls029; doi:10.1093/aobpla/pls029, available online at www.aobplants.oxfordjournals.org.

46. Singh VK, Singh A, Singh SP, Ellur RK, Singh D, Krishnan SG, Bhowmick PK, Nagarajan M, Vinod KK, Singh UD, Mohapatra T, Prabhu KV, Singh AK: Marker-assisted simultaneous but stepwise backcross breeding for pyramiding blast resistance genes Piz5 and Pi54 into an elite Basmati rice restorer line 'PRR78'. *Plant Breed* 2013, 132:486-495.

47. Sirithunya P, Tragoonrung S, Vanavichit A, Pa-in N, Vongsaprom C, Toojinda T:Quantitative trait loci associated with leaf and neck blast resistance in recombinant inbred line population of rice (*Oryza sativa*).m *DNA Res* 2002, 9:79-88.

48. Song WY, Wang GL, Chen LL, Kim HS, Pi LY, Holsten T, Gardner J, Wang B, Zhai WX, Zhu LH, Fauquet C, Ronald PC: A receptor kinase-like protein encoded by the rice disease resistance gene, Xa21. *Science* 1995, 270:1804-1806.

49. Sreewongchai T, Toojinda T, Thanintorn N, Kosawang C, Vanavichit A, Tharreau D, Sirithunya P: Development of elite indica rice lines with wide spectrum of resistance to Thai blast isolates by pyramiding multiple resistance QTLs. *Plant Breed* 2010, 129(2):176-180.

50. Suh JP, Roh JH, Cho YC, Han SS, Kim YG, Jena KK: The Pi40 gene for durable resistance to rice blast and molecular analysis of Pi40-advanced backcross breeding lines. *Phytopathology* 2009, 99:243-250.

51. Suh JP, Jeung JU, Noh TH, Cho YC, Park SH, Park HS, Shin MS, Kim CK, Jena KK:Development of breeding lines with three pyramided resistance genes that confer broad-spectrum bacterial blight resistance and their molecular analysis in rice. *Rice* 2013, 6:5.

52. Sun L, Su C, Wang C, Zai H, Wan J: Mapping of a major resistance gene to brown planthopper in the rice cultivar Rathu Heenati. *Breed Sci* 2005, 55:391-396.

53. Sun L, Liu Y, Jiang L, Su C, Wan J: Identification of quantitative trait loci associated with resistance to brown planthopper in the *indica* rice cultivar Col.5 Thailand. *Hereditas* 2006, 144:48-52.

54. Sundaram RM, Vishnupriya MR, Biradar SK, Laha GS, Reddy GA, Rani NS, Sarma NP, Sonti RV: Marker assisted introgression of bacterial blight resistance in Samba Mahsuri, an elite indica rice variety. *Euphytica* 2008, 160:411-422.

55. Tian Z, Qian Q, Liu Q, Yan M, Liu X, Yan C, Liu G, Gao Z, Tang S, Zeng D, Wang Y, Yu J, Gu M, Li J: Allelic diversities in rice starch biosynthesis lead to a diverse array of rice eating and cooking qualities. *PNAS* 2009, 106(51):21760-21765.

56. Tomita M: Introgression of green revolution sd1 gene into isogenic genome of rice super cultivar Koshihikari to create novel semidwarf cultivar 'Hikarishinseiki' (Koshihikari-sd1). *Field Crops Res* 2009, 114:173-181.

57. Umemoto T, Aoki N: Single-nucleotide polymorphisms in rice starch synthase IIa that alter starch gelatinization and starch association of the enzyme. *Funct Plant Bio* 2005, 32:763-768.

58. Umemoto T, Yano M, Satoh H, Shomura A, Nakamura Y: Mapping of a gene responsible for the difference in amylopectin structure between japonica-type and indica-type rice varieties. *Theor Appl Genet* 2002, 104:1-8.

59. Umemoto T, Aoki N, Lin HX, Nakamura Y, Inouchi N, Sato Y, Yano M, Hirabayashi H, Maruyama S: Natural variation in rice starch synthase IIa affects enzyme and starch properties. *Funct Plant Bio* 2004, 31:671-684.

60. Wang P, Xing YZ, Li ZK, Yu SB: Improving rice yield and quality by QTL pyramiding. *Mol Breeding* 2012, 29(4):903-913.

61. Watanabe T, Kitagawa H: Photosynthesis and translocation of assimilates in rice plants following phloem feeding by the planthopper *Nilaparvata lugens*(Homoptera: Delphacidae). *J Econ Entomol* 2000, 93:1192-1198.

62. Waters DL, Henry RJ, Reinke RF, Fitzgerald MA: Gelatinization temperature of rice explained by polymorphisms in starch synthase. *Plant Biotechnol J* 2006, 4(1):115-122.

63. Win KM, Korinsak S, Jantaboon J, Siangliw M, Lanceras-Siangliw J, Sirithunya P, Vanavichit A, Toojinda T: Breeding the Thai jasmine rice variety KDML105 for non-age-related broad-spectrum resistance to bacterial blight disease based on combined marker-assisted and phenotypic selection. *Field Crops Res* 2012, 137:186-194.

64. Wongsaprom C, Sirithunya P, Vanavichit A, Pantuwan G, Jongdee B, Sidhiwong N, Lanceras-Siangliw J, Toojinda T: Two introgressed quantitative trait loci confer a broad-spectrum resistance to blast

disease in the genetic background of the cultivar RD6 a Thai glutinous jasmine rice. *Field Crops Res* 2010, 119(2–3):245-251.

65. Xi ZY, He FH, Zeng RZ, Zhang ZM, Ding XH, Li WT, Zhang GQ: Development of a wide population of chromosome single-segment substitution lines in the genetic background of an elite cultivar of rice (*Oryza sativa* L.). *Genome* 2006, 49:476-484.
66. Yang Z, Sun X, Wang S, Zhang Q: Genetic and physical mapping of a new gene for bacterial blight resistance in rice. *Theor Appl Genet* 2003, 106:1467-1472.
67. Yi M, Nwe KT, Vanavichit A, Chai-arree W, Toojinda T: Marker assisted backcross breeding to improve cooking quality traits in Myanmar rice cultivar Manawthukha. *Field Crops Res* 2009, 113:178-186.
68. Yuan HY, Chen XP, Zhu LL, He GC: Identification of genes responsive to brown planthopper Nilaparvata lugens Stål (Homoptera: Delphacidae) feeding in rice. *Planta* 2005, 221:105-112.
69. Zong G, Wang AH, Wang L, Liang GH, Gu MH, Sang T, Han B: A pyramid breeding of eight grain-yield related quantitative trait loci based on marker-assistant and phenotype selection in rice (*Oryza sativa* L.).*J Genet Genom* 2012, 39:335-350.

Chapter 8

Gene Targeting and Genetic Transformation of Plants

Richard Mundembe[1]

[1]The Biotechnology Programme, Department of Agricultural and Food Sciences, Cape Peninsula University of Technology, Cape Town, South Africa

INTRODUCTION

A broad definition of gene targeting includes any method that can lead to permanent site-specific modification of the genome [1], preferably with predetermined outcomes. More specifically, gene targeting is the alteration of a specific DNA sequence in an endogenous gene at its original locus in the genome, and often refers to the conversion of the endogenous gene into a designed sequence [2]. Rapid developments in the field of gene targeting, and the potential of the technology to revolutionalise genomics and plant biotechnology in particular has led to the adoption of this broad definition, over earlier definitions such as that by [3] and [4] that restricted gene targeting to homologous recombination mechanisms.

While gene targeting does not necessarily lead to marker-free, vector backbone-free transformation, gene targeting certainly brings these desired outcomes of plant transformation research closer. Such marker-free, vector backbone-free plants will be truly and precisely engineered plants, and might actually be non-transgenic, depending on the source of the sequences used. Gene targeting in*Drosophila*, mice and yeast is now more or less routine [5]. Transgenic organisms for use in research are 'made-to-order' via gene targeting and are sold by commercial companies. Gene targeting in animals is accomplished via homologous recombination (HR). However, the same cannot be said of plants. Approaches adapted from gene targeting in yeast, insect and animal models have failed to give comparable results in plants mainly because the predominant mechanism of recombination in somatic cells of plants is not HR, but is non-homologous end joining, NHEJ, also known as illegitimate recombination [6].

Double-stranded breaks in plant genomic DNA are repaired either via HR or NHEJ [7]. Homologous recombination mechanisms involve linkage of DNA fragments to regions of identical sequence, such as the other member of the homologous partner, as template for accurate repair of the double stranded break. This mechanism is therefore only functional in the S/G2 phase of the cell cycle. Non-homologous end-joining mechanisms of recombination however are functional in all phases of the life cycle and do not require significant homology to join two fragments of a broken DNA molecule. While HR has been very successful in insects and animals [8], it has remained unavailable for the manipulation of plant transformation; it is NHEJ that is useful for plant transformation.

Occurrence of single-stranded breaks on a DNA molecule does not normally pose a challenge to the plant genome because these can be repaired by ligation without change to the nucleotide sequence. Faithful strand replacement or nick translation may take place starting at the single-strand break, again with no changes to the nucleotide sequence.

Double-stranded breaks, however, have dire consequences if not repaired, or if repaired incorrectly. A double-stranded break effectively results in two fragments of the chromosome, and only one of the fragments might have a centromere to enable separation after cell division; the other fragment might be 'lost'. Also, if unprotected, the double-stranded breaks are exposed to the exonucleases of the cell and may be misconstrued as foreign and will therefore be degraded.

Living cells therefore need efficient mechanisms for detecting chromosomal double-stranded breaks and initiation of appropriate repair mechanisms for replication to be successful. The repair of double-stranded breaks takes place by one of two main pathways for double-stranded break repair: the HR pathway or the NHEJ pathway, or both. Coincidentally, these are the two mechanisms by which exogenous DNA may also integrate into a host genome [7].

HOMOLOGOUS AND NON-HOMOLOGOUS RECOMBINATION

Recombination evolved in nature to repair DNA damage that may occur during the cell cycle, and to generate diversity through meiotic recombination of genetic material which in turn has enabled sexually-reproducing eukaryotes to become extremely adaptive to their ever-changing environment and is partly responsible for their success on earth.

Homologous Recombination

In homologous recombination (HR) a long and extended region of homology such as that found between sister chromatids is required for the two DNA molecules to line up adjacent to each other. There are many variations to this pathway, but the basics of two popular models are illustrated inFigure 1. A cellular protein, Spo I, may induce double-stranded breaks in the chromosome. These double-stranded breaks are repaired exclusively by HR using one of several possible homologous matrices: copied from elsewhere in the genome (ectopic HR), copied from the homologue (allelic HR), or copied from the same chromosome (intra-chromosomal HR) [6, 9].

Ectopic HR is a minor pathway, and was reported to be responsible for the repair of only one in 10 000 double-stranded breaks. In some of the cases, both homologous and non-homologous end-joining mechanisms were involved in repairing different ends of the same double-stranded break [6, 10]. Of the possible ectopic recombination models, the synthesis-dependent strand annealing (SDSA) model is the one that is conservative and is consistent with these observations. Figure 1(a) below illustrates this model.

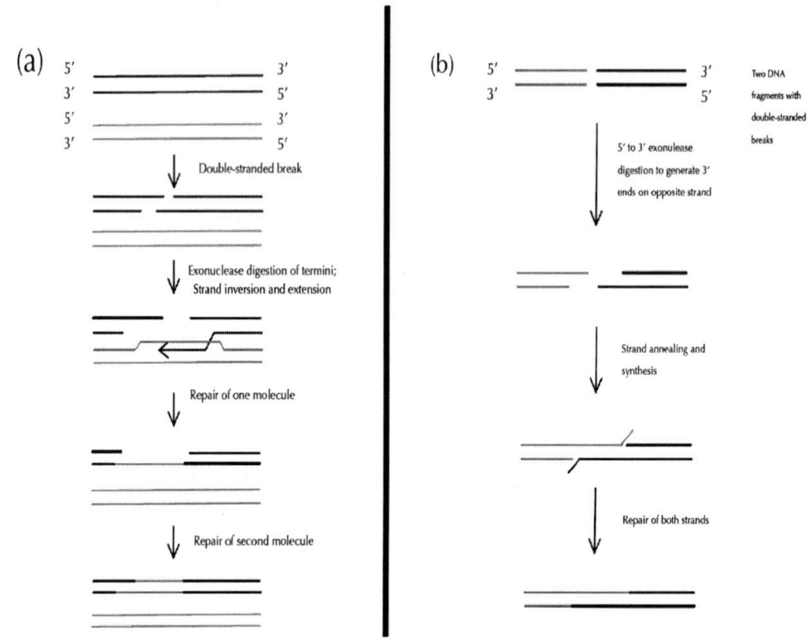

Figure 1: Models for double-stranded break repair mechanisms. (a) The synthesis-dependent strand-annealing (SDSA) for ectopic recombination. (b) The single-strand annealing (SSA) model. Polarity of the DNA molecules is shown on the first set of molecules only for simplicity. Modified from [7].

This model predicts that double-stranded break repair is not accompanied by crossing-over. Also, both perfect integrations into target sites by homologous recombination and imperfect integrations by HR are possible on one end of the target site. Integration by NHEJ is possible on the other end of the same double-stranded break of transgene, as well as ectopic integrations elsewhere in the genome, after copying of transgene sequences.

Allelic HR occurs during meiosis, to repair double-stranded breaks using sequences of the homologues in a process that involves formation of Holliday junctions to resolve into the crossover or gene conversion products. Allelic HR is not significant in somatic cells but is the classic HR that occurs in meiotic cells. In nature this essential process takes place during meiosis I to result in recombination for sexually reproducing species. Extensive lengths of homology (several hundreds or thousands of nucleotides) are required for this process, and ensures that recombination takes place between sister chromatids.

Intrachromosomal HR utilizes sequences close to the double-stranded break, on the same chromosome or on the sister chromatid (in G_2 stage only) as a matrix for repair. This can result in deletion as predicted by single-strand annealing (SSA) model (Figure 1b) or gene conversion as predicted by the conservative SDSA model depending on the structure of the chromosomal locus [11]. The SSA pathway was shown to be five time more efficient than the SDSA pathway [12]. SSA-like pathways have also been described for NHEJ.

Non-Homologous End-Joining Recombination

The second pathway is non-homologous end-joining (NHEJ) pathway, also known as illegitimate recombination. It also requires some homology, albeit much reduced. This limited homology is required at the ends of the DNA strands on which the double-stranded breaks occur. The double-stranded breaks can be sticky ends or blunt ends. The homology present within the sticky ends may be sufficient for this mechanism, and the properly aligned ends will be ligated together. For blunt ends, binding of a specific protein complex, such as the Ku complex in mammalian cells, to the broken ends of the DNA limits nucleolytic degradation, and unlike HR repair, prevents exposure of single-stranded regions [8]. The bound protein may also function directly or indirectly to bring the DNA ends together for processing and ligation. Alignment of the termini by complementary micro-homologies of 1 – 4 nucleotides is usually required. The process might also require either limited unwinding or limited exonucleolytic digestion to expose the ends for alignment, and DNA polymerase to fill-in gaps. Single-stranded deletion of short segments at the 5′-end may expose single-stranded regions that will be used to search for homologies in the other DNA fragment, which will then form the basis of the alignment and repair [8]. The process of NHEJ is illustrated in Figure 2. The arrangement of chromosomal DNA into loops attached to a matrix that restricts the mobility of DNA promotes the re-joining of previously linked DNA ends [8].

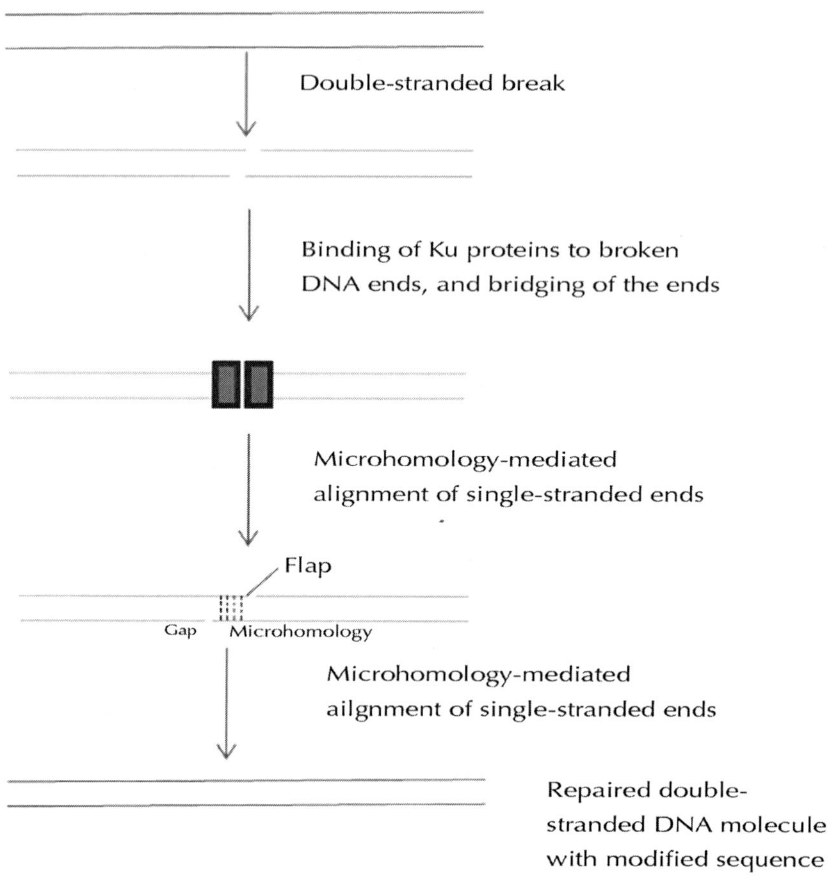

Figure 2: Model for repair of double-stranded breaks by non-homologous end-joining. Adapted from [8].

NHEJ is the predominant pathway for double-stranded break repair in somatic cells of higher eukaryote, including plants. Simple ligation will result in junctions with no homology. Short stretches of homology may be a result of SSA-like mechanisms [13], while longer stretches might be from an SDSA copying of ectopic chromosomal DNA into the break [14].

NHEJ is also the mechanism by which transgene integration occurs following either *Agrobacterium*-mediated or direct transformation of plant cells. The integration sites are generally random, but transcriptionally active sites seem to be preferred.

When we consider the evolution of gene targeting research, HR pathways were initially considered the only route with potential to achieve this because of high levels of fidelity observed in HR during meiosis. The levels of homology involved in meiotic recombination are large and would make this approach unworkable for routine plant genetic engineering. The extent of homology required is extensive, and may elongate the transgenes required in plant transformation to impractical levels. Induction of double-stranded breaks on the DNA by exposure to X-rays or by transposon activity was shown to increase HR [15, 16]. Site-specific recombination systems therefore became a potential route to achieving gene targeting by HR, since they can introduce double-stranded breaks in DNA, and repair these in via an HR mechanism that utilizes shorter homologies.

The objective of many plant transformation research groups is to study genomics and generate improved crops. While transgenic plants produced for genomics study have little regulatory requirements since they are for contained use, transgenic plants for general release have to comply with governmental regulations and must also meet consumer acceptance. Gene targeting will make it easier for genetically modified plants to meet these requirements. A strategy for gene targeting that has been explored extensively by researchers is that of site-specific recombination.

Site-specific recombination systems consist of a recombinase and donor sites. The recombinase is a protein that mediates a recombination reaction between a target site characterized by particular target sequence for that protein, and the donor site, also with a characteristic nucleotide sequence. In general, the results of the ensuing recombination reaction are excision, integration or inversion.

The site-specific recombination systems that can be utilized for gene targeting include the tyrosine family recombination system, the serine family recombination system and the newly developed hybrid system consisting of zinc finger DNA sequence recognition motifs in combination with a rare-cutting restriction endonuclease. Each of these systems will now be considered in turn, and the potential to contribute to gene targeting discussed.

TYROSINE FAMILY RECOMBINATION SYSTEMS

The tyrosine family recombination systems include the Cre/*lox P*, FLP/*FRP*, λ integrase and variations thereof [17]. Also known as integrases, they use the hydroxyl group of the catalytic tyrosine for a nucleophilic attack on the phosphodiester bond of the target DNA, and function through a Holliday junction intermediate [17, 18]. Their function depends on the cofactors supplied.

Cre and FLP recombinases are the most popular members of the integrase family because they are simple and unrestrictive, requiring no auxillary factors other than their recombinase monomers and their cognate targets. Cre recombinase recombines 34 bp *lox P* sites in the absence of accessory proteins or auxillary DNA sequences [17, 19]. The FLP target site has been trimmed from the original 599 bp in the 2 μm yeast plasmid to 30 bp in *FRT* sites [17]. The wild-type *FRT* and *lox P* sites are unchanged by the recombination reaction, making the reaction reversible; and there are many different possible recombination intermediates in each case. This has however made it difficult to utilize the tyrosine family recombinase systems more extensively in vector construction, while the irreversible λ integrase is more popularly used in vectors [17].

SERINE FAMILY RECOMBINASE SYSTEMS

The Serine family recombinase systems such as □C31, Hin and Gin [17] are also known as the resolvases or invertases. They have a conserved serine residue that is used to create the covalent link between the recombinase and the DNA target site [18]. Serine family recombinases initiate strand-exchange by making double-stranded breaks at two sites in the DNA molecules. Each site of the double-stranded break is associated with a dimer of the recombinase, and the two dimers will come together bringing the two broken ends together and forming an active tetramer in a process that is elaborately controlled [20].

The general scheme for using site-specific recombination systems in gene targeting involves, first, the genetic engineering of the recombination target sites into the particular genomic location of the plant to be transformed. This can be achieved by standard transformation procedures followed by screening to identify transformation events in 'acceptable' locations. Transposon tagging has also been used with the recombination target sites incorporated within the transposon.

The second requirement is that the incoming transgene should have unique DNA sequences that constitute the donor sites. Finally, there should be a mechanism for expression or introduction of the recombinase, to mediate the recombination reaction between donor and target sites. In this scheme, a second transformation experiment targets the genes into which the recombinase target sequences were integrated by the first transformation experiment.

Both *Agrobacterium*-mediated and direct gene transfer (bombardment, electroporation and PEG-mediated transformation) have been used for the initial transformation to introduce target and donor sites. The recombinase may be expressed constitutively, transiently or may be induced. Recombinase expression as well as stability of the transgene may vary [21].

These approaches were based on the need to improve homologous recombination at the target site. In these approaches, homology is limited to target and donor site compatibility for the particular recombinase being considered. With elegant engineering, site-specific recombination systems can be used to remove marker genes from transgenic plants before their commercialization. But the process is far from routine. Also, the footprint that remains on the chromosome is associated with genetic instability. The search for a better system continues, and that is why zinc finger nucleases are being considered.

ZINC FINGER NUCLEASE AND GENE TARGETING

Zinc finger nucleases (ZFN) are artificial restriction endonucleases composed of a fusion between an artificial Cys_2His_2 zinc finger protein DNA binding domain and the cleavage domain of the *Fok* I endonuclease. The sequence-specific DNA binding domain of zinc fingers could be engineered to recognize a variety of specific DNA nucleotide sequences of the researcher's choice [22]. While the*Fok* I endonuclease activity is non-specific, the enzyme only functions when it forms a dimer, whose assembly will be guided by proper alignment of the two zinc finger monomers at the target site. Assembly of the ZFN therefore enables site-specific cleavage [3, 23]. The two zinc finger monomers are usually designed to flank a 5 – 6 bp long target sequence, within which *Fok* I cleavage will occur. The zinc finger domain itself is composed of 3 – 4 individual fingers, each of which recognizes 3 bp sequences [1]. Overall, a unique sequence of about 24 bp is specifically recognized, and this is large enough to be unique in most genomes.

The most common forms of the ZFN recognition sites are $(NNY)_3N_6(RNN)_3$, of which $(NNC)_3N_6(GNN)_3$ has been extensively studied [23, 24]. The double-stranded breaks will significantly increase integration of DNA into the target site by HR by up to 100 times in plants [1]. But even then, double-stranded breaks induced by restrictions endonucleases or transposons have been shown to be predominantly repaired by NHEJ, often accompanied by some level of mutagenesis [1, 7]. A high proportion of the double-stranded breaks will therefore be repaired by NHEJ, since it is the predominant repair mechanism in plants.

Once the double-stranded break is made, early approaches were to try and increase the chances of their being repaired by HR, over the more predominant NHEJ. The approach has not been very successful. Research efforts should rather focus on ensuring that the repair by NHEJ does not mutate the nucleotide sequence of the target gene in an undesirable manner.

There have been attempts to increase the chances of HR after inducing double-stranded breaks with ZFN. For example, use of ZFN in combination with recombinases and chromatin re-modeling proteins, this system increases both targeting precision and transformation efficiencies by HR. Further development of the system should optimise removal or exclusion of marker and reporter genes as well as vector backbone sequences.

Gene targeting using ZFN was first demonstrated for the yellow locus in *Drosophila* [25]. The approach has since become standard for many animal species having been demonstrated even in humans

[26]. In higher plants, the technology lagged behind, but has shown a lot of promise with for instance the use of a novel TRV-based vector to achieve non-transgenic genome modification in plant cells [27]. ZFN may also be used for gene deletion [28], and removal of marker genes. Genetically engineered plants released into the environment should not have unnecessary transgene sequences.

Failure to increase HR in plants does not mean that all is lost. In fact, maybe plant transformation efforts may well benefit from NHEJ, which is the predominant mechanism of recombination in plants cells anyway. If one considers for instance a scenario where one needs to disrupt an endogenous gene whose phenotype is easily assayed for. Transient expression of a ZFN that targets the gene should introduce double-stranded breaks in the gene, and most of the breaks will be repaired by NHEJ. Errors introduced during the repair process should inactivate the gene. The precisely engineered plant you get is non-transgenic because there are no foreign sequences integrated into its genome, but the genome would have been elegantly edited! Sequencing of the edited gene should be used to confirm and characterize the mutation. Marton and coworkers reported on a successful experiment with this approach using a disarmed *Tobacco rattle virus* (TRV) vector to deliver the ZFN to cells of intact tobacco and petunia plants [27]. The mutations that were induced were stable and heritable.

There are many other possibilities for gene targeting in plants. For instance, the efficacy of oligonucleotide-directed plant gene targeting has been demonstrated, again with the possibility of the plants being considered non-transgenic [29].

PROSPECTS FOR FURTHER DEVELOPMENT

The efficacy of gene targeting in plants has now been demonstrated, and genetically engineered plants using this technology are being developed. These plants are expected to be low copy number, reflecting on the target gene, and in genomic locations that correspond to the natural locations of the targets. Gene targeting approaches utilize the vast amount of genomic data that is now readily available in databases and can be correlated with the stability of the modification introduced at particular genomic sites. This would be ideal to enhance agricultural attributes of crop, for instance by increasing the expression of a desired product or shutting down a competing or undesirable pathway. With the levels of precision and true engineering that comes with gene targeting, the dependency on reporter genes and even marker genes is reduced. New and elegant ways of delivering DNA to plant cells, such as oligonucleotides and minimal cassettes will enable plant transformation without the use of plant transformation vectors whose backbones are notorious for integrating into the plant genome. Marker-free, vector backbone-free precisely engineered agricultural crops are what farmers, consumers and the environment need.

CONCLUSIONS

Gene targeting technology in plants has come a long way, and several alternative approaches to gene targeting have been evaluated. It is now possible and desirable for new plant transformation experiments to give some consideration as to which region of the genome they would want to target, and also give special consideration to reporter genes, marker genes and vector backbone sequences that might be associated with the experiment. It is hoped that the dream of reporter-free, marker-free, vector backbone-free truly and precisely engineered plants will soon be a reality.

REFERENCES

1. D Weinthal, A Tovkach, V Zeevi, and T Tzfira, 2010Genome editing in plant cells by zinc finger nucleases. Trends in Plant Science 15 308 321
2. S Iida, and R Terada, 2005Modification of endogenous natural genes by gene targeting in rice and other higher plants. Plant Molecular Biology 59 205 219
3. S Durai, M Mani, K Kandavelou, J Wu, M. H Porteus, and S Chandrasegaran, 2005Zinc finger nucleases: custom-designed molecular scissors for genome engineering of plant and mammalian cells. Nucleic Acids Research 33 18 5978 5990

4. R Terada, Y Johzuka-hisatomi, M Saitoh, H Asao, and S Iida, 2007Gene targeting by homologous recombination as a biotechnological tool for rice functional genomics. Plant Physiology 144 846 856
5. M. J Evans, O Smithies, and M. R Capecchi, 2001Mouse gene targeting. Nat Med 7 1081 1090
6. H Puchta, 1999DSB-induced recombination between ectopic homologous sequences in somatic plant cells. Genetics 152 1173 1181
7. H Putcha, 2005The repair of double-strand breaks in plants: mechanisms and consequences for genome evolution. Journal of Experimental Botany 56 409 1 14
8. M. L Hefferin, and A. E Tomkinson, 2005Mechanism of DNA double-strand break repair by non-homologous end joining. DNA Repair 4 639 648
9. S Keeney, 2001Mechanism and control of meiotic recombination initiation. Current Topics in Developmental Biology 52 1 53
10. G Shaley, and A. A Levy, 1997The maize transposable genetic element Ac induces recombination between donor site and a homologous ectopic sequence. Genetics 146 1143 1151
11. J Fishman-lobell, N Rudin, and J. E Haber, 1992Two alternative pathways of double-strand break repair that are kinetically separable and independently modulated. Molecular and Cellular Biology 12 1292 1303
12. N Orel, A Kirik, and H Puchta, 2003Different pathways of homologous recombination are used for repair of double-stranded breaks within tandemly arranged sequences in the plant genome. The Plant Journal 35 604 712
13. A. L Nicolas, P. L Munz, and C. S Young, 1995A modified single-strand annealing model best explains the joining of DNA double-strand breaks in mammalian cells and cell extracts. Nucleic Acids Research 23 1036 1043
14. S Salomon, and H Puchta, 1998Capture of genomic and T-DNA sequences during double-strand break repair in somatic plant cells. EMBO Journal 17 6086 6095

15. J Tovar, and C Lichtenstein, 1992Somatic and meiotic chromosomal recombination between inverted duplications in transgenic tobacco plants. The Plant Cell 4 319 332
16. Y. L Xiao, and T Peterson, 2000Interchromosomal homologous recombination in Arabidopsis induced by a maize transposon. Molecular and General Genetics 263 22 29
17. L. A Lyznik, W Gordon-kamm, H Gao, and C Scelonge, 2007Application of site-specific recombination systems for targeted modification of plant genomes. Transgenic Plant Research 1 1 1 9
18. N. D Grindley, K. L Whiteson, and P. A Rice, 2006Mechanisms of site-specific recombination. Annual Reviews in Biochemistry 75 567 605
19. D. N Gopaul, F Guo, and G. D Van Duyne, 1998Structure of the Holliday junction intermediate in Cre-loxP site-specific recombination. The EMBO Journal 17 14 4175 4187
20. S. W Marshall, M. R Boocock, F. J Olorunniji, and S. J Rowland, Intermediates in serine recombinase-mediated site-specific recombination. Biochemical Society Transections 39 2 617 622
21. S Kumar, M Franco, and G. C Allen, 2006Gene targeting: Development of novel systems for genome engineering in plants. Floriculture, Ornamental and Plant Biotechnology Volume IV, Chapter 8, 84 98Global Science Books.
22. F Fauser, N Roth, M Pacher, G Ilg, Sanchez-Fernandez, Biesgen, C. and Puchta, H. (2012In plant gene targeting. PNAS 109 19 7535 7540
23. D Carroll, J. J Morton, K. J Beumer, and D. J Segal, 2006Design, construction and in vitro testing of zinc finger nucleases. Nature Protocols 1 1329 1341
24. Q Liu, Z. Q Xia, X Zhong, and C. C Case, 2002Validated zinc finger protein designs for all 16 GNN DNA triplet targets. Journal of Biological Chemistry 277 3850 3856
25. M Bibikova, D Carroll, D. J Segal, J. K Trautman, J Smith, Y-G Kim, and S Chandrasegaran, 2001Stimulation of homologous recombination through targeted cleavage by chimeric nucleases. Molecular and Cell Biology 21 289 297

26. M. H Porteus, and D Carroll, 2005Gene targeting using zinc finger nucleases. Nature Biotechnology 23 967 973

27. I Marton, A Zuker, E Shklarman, V Zeevi, A Tovkach, S Roffe, M Ovadis, T Tzfira, and A Vainstein, 2010Non-transgenic genome modification in plant cells. Plant Physiology154 1079 1087

28. J. F Petolino, A Worden, K Curlee, J Connell, T. L. S Moynahan, C Larsen, and S Russell, 2010Plant Molecular Biology 73 617 628

29. T. J Oh, and G. D May, 2001Oligonucleotide-directed gene targeting. Current Opinion in Biotechnology12: 169 172.

Citations

CHAPTER 1

Joel P Tully, Aubrey E Hill, Hadia MR Ahmed, Ryan Whitley, Anthony Skjellum, and M Shahid Mukhtar, Expression-Based Network Biology Identifies Immune-Related Functional Modules Involved in Plant Defense, doi:10.1186/1471-2164-15-421.

CHAPTER 2

Loren A Honaas, Eric K Wafula, Zhenzhen Yang, Joshua P Der, Norman J Wickett,Naomi S Altman, Christopher G Taylor, John I Yoder, Michael P Timko, James H Westwood, and Claude W dePamphilis, Functional Genomics of a Generalist Parasitic Plant: Laser Microdissection of Host-Parasite Interface Reveals Host-Specific Patterns of Parasite Gene Expression, doi:10.1186/1471-2229-13-9.

CHAPTER 3

Zi Shi, Siela N Maximova, Yi Liu, Joseph Verica, and Mark J Guiltinan, Functional Analysis of the *Theobroma Cacao NPR1* Gene in *Arabidopsis,* doi:10.1186/1471-2229-10-248.

CHAPTER 4

Vinzenz Hofferek, Amelie Mendrinna, Nicole Gaude, Franziska Krajinski, and Emanuel A Devers, Mir171h Restricts Root Symbioses and Shows Like its Target*NSP2* a Complex Transcriptional Regulation in *Medicago Truncatula,* doi:10.1186/s12870-014-0199-1.

CHAPTER 5

Jin-Min Tu, Ming-Chung Chang, Lynn LH Huang, Ching-Dong Chang, Hao-Jen Huang, Ruey-Hua Lee, and Ching-Chun Chang, The blue fluorescent protein from Vibrio vulnificus CKM-1 is a useful reporter for plant research, doi:10.1186/s40529-014-0079-x.

CHAPTER 6

Zhaowen Mo, Wu Li, Shenggang Pan, Timothy L Fitzgerald, Feng Xiao, Yongjian Tang, Yilei Wang, Meiyang Duan, Hua Tian, and Xiangru Tang, Shading during the grain filling period increases 2-acetyl-1-pyrroline content in fragrant rice, doi: 10.1186/s12284-015-0040-y.

CHAPTER 7

Siriphat Ruengphayak, Ekawat Chaichumpoo, Supaporn Phromphan, Wintai Kamolsukyunyong, Wissarut Sukhaket, Ekapol Phuvanartnarubal, Siripar Korinsak, Siriporn Korinsak, and Apichart Vanavichit, Pseudo-Backcrossing Design for Rapidly Pyramiding Multiple Traits Into a Preferential Rice Variety, doi:10.1186/s12284-014-0035-0.

CHAPTER 8

Richard Mundembe (2013). Gene Targeting and Genetic Transformation of Plants, Genetic Engineering, Prof. Idah Sithole-Niang (Ed.), ISBN: 978-953-51-1099-6, InTech, DOI: 10.5772/56335.

Index

D

E

F

G

H

I

J

L

M

N

O